Springer Series in
SOLID-STATE SCIENCES

Springer Series in
SOLID-STATE SCIENCES

Series Editors:
M. Cardona   P. Fulde   K. von Klitzing   R. Merlin   H.-J. Queisser   H. Störmer

The Springer Series in Solid-State Sciences consists of fundamental scientific books prepared by leading researchers in the foield. They strive to communicate, in a systematic and comprehensive way, the basic principles as well as new developments in theoretical and experimental solid-state physics.

Peter Mohn

Magnetism in the Solid State

An Introduction

Corrected Second Printing

With 72 Figures and 7 Tables

 Springer

Prof. Dr. Peter Mohn
Vienna University of Technology
Center for Computational Materials Science
Getreidemarkt 9/134
1060 Vienna, Austria
E-mail: phm@cms.tuwien.ac.at

Series Editors:

Professor Dr., Dres. h. c. Manuel Cardona
Professor Dr., Dres. h. c. Peter Fulde*
Professor Dr., Dres. h. c. Klaus von Klitzing
Professor Dr., Dres. h. c. Hans-Joachim Queisser

Max-Planck-Institut für Festkörperforschung, Heisenbergstrasse 1, 70569 Stuttgart, Germany
* Max-Planck-Institut für Physik komplexer Systeme, Nöthnitzer Strasse 38
  01187 Dresden, Germany

Professor Dr. Roberto Merlin

Department of Physics, 5000 East University, University of Michigan
Ann Arbor, MI 48109-1120, USA

Professor Dr. Horst Störmer

Dept. Phys. and Dept. Appl. Physics, Columbia University, New York, NY 10027 and
Bell Labs., Lucent Technologies, Murray Hill, NJ 07974, USA

ISSN 0171-1873
Corrected 2nd Printing 2006
ISBN-10 3-540-29384-1   Springer Berlin Heidelberg New York
ISBN-13 978-3-540-29384-2   Springer Berlin Heidelberg New York
ISBN 3-540-43183-7   1st Ed. Springer-Verlag Berlin Heidelberg New York

Library of Congress Control Number: 2005933346

Springer is a part of Springer Science+Business Media
springeronline.com

Typesetting by the authors
Cover concept: eStudio Calamar Steinen
Cover production: *design & production* GmbH, Heidelberg
Production and data conversion: LE-TeX Jelonek, Schmidt & Vöckler GbR, Leipzig

Printed on acid-free paper   SPIN: 11568209   57/3141/YL - 5 4 3 2 1 0

To the memory of E. Peter Wohlfarth, a great magnetician and humanist

Preface

During 1986 and 1987, when I was working at the Imperial College of Science and Technology in London with E.P. Wohlfarth, I was asked to prepare some lecture notes following his lecture *"Magnetic properties of metals and alloys"*. Peter Wohlfarth had planned to use these lecture notes in a book about itinerant magnetism he wanted us to write together. Due to his untimely death this project was never realized.

Back in Vienna I started to write up the lecture notes for a graduate course, which to some extent was based on these original lectures but in many respects has been modernized by including more recent results of band structure calculations and band theoretical results. During a two month visit of Melbourne University in 1993, I met G. Fletcher from Monash University (Melbourne) who kindly offered to critically read the manuscript. His comments are highly appreciated.

The present greatly enlarged version was mainly written during a sabbatical at the university of Uppsala during 2000. I am grateful to Börje Johansson for making this possible and to Clas Persson for reading the manuscript and providing me with lots of useful comments.

The aim of the book is to present a largely phenomenological introduction to the field of solid state magnetism at a relatively elementary level. The two basic concepts of magnetism in solids namely the localized and the delocalized description are presented as the extreme approaches. The true nature of magnetism lies, as often in life, somewhere in between, sometimes showing a tendency towards the more localized side, sometimes tending to the delocalized side. It is perhaps this mixing of concepts which makes magnetism appear complicated and difficult. Another source of confusion is the different language used by theoreticians and experimentalists. I have tried very hard to clarify these rather more semantic problems and to use a uniform nomenclature throughout the book. It is my belief and my experience that the approach presented here provides a useful introduction not only for the physicist, but also for the interested reader coming from fields like chemistry, electrical engineering or even geo-sciences. The mathematical concepts used are kept rather simple and hardly ever go beyond an undergraduate course in mathematics for physicists, chemists or engineering. Since the book emerged from a lecture course I have given at Vienna University of Technology for the last 15 years, the chapters in the book are not completely self-contained. The first-time reader is thus advised to read the chapters in the sequence that they appear in the book. It is my sincere hope that after having read this book the reader will agree that for once the Encyclopedia Brittanica is in error when it states *Few subjects in science are more difficult to understand than magnetism,* (Encyclopedia Brittanica, 15th edition 1989).

The present book does not attempt to cover the whole field of solid state magnetism, but tries to provide an overview by selecting special topics. The idea is to create an interest in this fascinating field in which quantum mechanics, thermodynamics and computer simulations join forces to explain "Magnetism in the Solid State".

Vienna,   *Peter Mohn*
June 2002

Since this book has been so well received by the scientific community that the first printing has been sold within three years, I was asked by the publishers to produce an updated version for a second printing. I am grateful to all colleagues (mainly students) who reported typos, ambiguous or unclear formulations etc. I considered all of them seriously and thus made a number of changes, which I hope will improve the text.

Vienna,   *Peter Mohn*
September 2005

Contents

1. A Historical Introduction

Already in archaic times it was known that a type of stone which was found close to a place called Magnesia in northern Greece attracts iron. The Greek philosopher and mathematician Thales of Milet (about 625-564 b.c.) [1] even attributed a soul and thus life to this attracting material. The first known application of magnetism is the compass as it was used by the Chinese [2]. The oldest description of a compass is found in the book *Meng Chhi Pi Than* (from the year 1086) of the author Shen Kua who not only describes the compass needle to point to the south direction but also refers to a slight easterly deviation meaning the declination of the compass. Although only described therein for the first time, it seems probable that the compass was already used in the 7th and 8th century since during this time the natural magnet lodestone (formed by the mineral magnetite; Fe_3O_4) was already replaced by iron needles. The Chinese used the compass mainly for terrestrial navigation; only at the end of the 12th century did it also appear on ships.

In Europe the compass was first mentioned in 1187 by the Englishman A. Neckam [3] in his works *De utensilibus* and *De naturis rerum*. Other sources about the use of magnetic needles can be found in a poem by G. de Provins [4] and in a letter of the crusader P. de Mericourt in his *Epistola de magnete* [5]. It is interesting that, although it was well known that magnets only attract iron and iron rich metals, the medieval literature also reports tales about magnetic mountains or islands whose magnetism was said to be able to remove the copper or bronze nails out of boats. It took until 1600 for Gilbert [6] to notice that the use of the right kind of iron is necessary to produce strong magnets. Figure 1.1 shows how a piece of iron can be magnetized by coldworking in the earth's magnetic field.

During the next centuries people tried to produce stronger magnets; in particular Knight [7] succeeded in building a magnetic "magazine" which was strong enough to reverse the magnetization of any other known magnet put into it. Also in the medical science of the 18th century the use of magnets came into fashion. In particular the German physician Mesmer claimed to be highly successful in curing various diseases by magnetizing the patient. In 1775 he published his results and claimed that his discovery of what he called *Magnetismus animalis* is of medical relevance. His influence was so strong that his name even entered the English language and is still present in the

Fig. 1.1. Coldworking of a permanent magnet in the direction of the earth's magnetic field (septentrio=north, auster=south) taken from Gilbert [6]

word *mesmerizing*. In a rather satirical way Lorenzo da Ponte comments on the Mesmer hysteria of his time in the textbook to Mozart's opera *Cosi fan tutte* [8] where in one scene two young men, who pretend to have poisoned themselves, are brought back to life by the application of a giant horseshoe magnet. This is not the only lyrical approach to magnetism. A rather different one can be found in the operetta *Patience* by Gilbert and Sullivan where W.S. Gilbert wrote in 1881:[1]

A magnet hung in a hardware shop,
And all around him was loving crop
Of scissors and needles, nails and knives,
Offering love for all their lives;
But for iron the magnet felt no whim,

---

[1] I am grateful to G. Fletcher for bringing this "magnetism chant" to my attention. It has also been referred to by J.H. van Vleck [9].

Though he charmed iron, it charmed not him;
From needles and nails and knives he'd turn,
For he'd set his love on a Silver Churn!
... But this magnetic, peripathetic
Lover he lived to learn,
By no endeavour
Can magnet ever
Attract a Silver Churn!

The effect of electric currents on compass needles was discovered by Oersted in 1820. As early as in 1831 Faraday formulated the induction principle which finally gave rise to a new scientific discipline: electromagnetism. The first electromagnets were built by Sturgeon [10] whose work started also a new and more systematic search for improving magnets. With the development of exact sciences the need for well defined stable magnetic fields also grew and reached its climax with the work of F. Bitter during the 1930s who succeeded in producing fields up to 15T in a bore of 5cm diameter. He developed segmented coils (which could tolerate the enormous mechanical stress) which were water cooled and had a power consumption of about 5MW. This line of experimental setup soon came to an end not only because of the enormous power consumption and cooling problems, but also because of the development of superconducting magnets. Although the mechanical stress problems remain, coils from technical superconductors like Nb-Ti and Nb$_3$Sn with upper critical fields of 15 and 23T, respectively, are readily available. The highest fields produced in the laboratory come from pulsed magnets, where a capacitance battery is discharged through a coil. Experiments are then performed at the peak flux usually for periods less than 1ms. These appliances reach up to 120T. A further increase can be achieved by implosion coils, where at the time of the main flux from a pulsed current, an implosion charge is ignited which contracts the coil area and thus increases the flux. Due to the high cost of these "self destroying" experiments their application is rather limited.

The development of "microscopic" models of the magnetic properties of free atoms, molecules and (much later) solids, started in the late 19th and early 20th century. It required the formulation of Maxwell's electrodynamics and the ideas of Boltzmann's statistical thermodynamics to treat the properties of ensembles of electric and magnetic carriers. If one assumes that a molecule has a magnetic moment of magnitude μ, the paramagnetic susceptibility χ_{para} of an ensemble of such molecules will be given by

$$\chi_{\mathrm{para}} = \frac{N\mu^2}{3k_{\mathrm{B}}T} \quad . \tag{1.1}$$

These persistent magnetic moments are thought to be represented by tiny permanent magnets, which themselves are supposed to be rigid and hence incapable of induced polarization. At $T = 0$K these elementary magnets will all be aligned parallel to an applied field. At finite T, temperature agitation will reduce the average number of aligned moments in the way expressed in (1.1). The linear temperature dependence of the inverse susceptibility was noticed experimentally by P. Curie [11] and was later derived theoretically by Langevin [12]. Equation (1.1) is in fact known as Curie's law and describes the susceptibility of all systems in the classical limit (high temperature).

While the Langevin paramagnetism always yields a positive contribution to the susceptibility there must also exist a different mechanism which leads to a diamagnetic behavior. Langevin also showed that induced polarization leads to a diamagnetic response by assuming that an applied field induces an additional electric current in the electron system. Due to the classical induction law, the magnetic field produced by this current is opposed to the direction of the applied field (Lenz's rule) and thus weakens it, which in turn leads to a negative value of the susceptibility. On calculating the statistical average of the two-dimensional projection over a three-dimensional motion Langevin missed out a factor of 2, which was only later corrected by Pauli [13] so the diamagnetic susceptibility χ_{dia} reads

$$\chi_{\text{dia}} = -\frac{Ne^2}{6mc^2} \sum_i \langle r_i^2 \rangle \quad . \tag{1.2}$$

In (1.2) the quantity $\langle r_i^2 \rangle$ is the average radius of the motion of electron i. A comparison of the para- and diamagnetic susceptibilities shows the latter one to be usually much smaller. In fact, for most systems with open electronic shells the diamagnetic part is negligible. Only in cases where all electron shells are filled and the paramagnetic contribution (ideally) becomes zero, can a net diamagnetic susceptibility be observed (e.g. copper, noble-gases).

At this point of the development of a theory of magnetism there existed a beautifully simple model for the understanding of both para- and diamagnetism which seemed to be based on purely classical physics. The drawback came when in 1919 Miss van Leeuwen, a Ph.D. student of Niels Bohr, demonstrated that classical Boltzmann statistics applied rigorously to any dynamical system must lead to a zero susceptibility [14]. The proof of the theorem, which is also referred to by van Vleck [15], is most elucidating for the understanding of magnetism and is briefly reviewed here.

Unambiguously one can assume that any magnetic moment, which has to be related to an angular momentum of a charged particle, can be written as a linear function

$$m_z = \sum_{k=1}^{f} a_k \dot{q}_k \quad , \tag{1.3}$$

of the generalized velocities $q_1 ... q_f$. This assumption is particularly clear in Cartesian coordinates where the magnetic moment (e.g. m_z) is given as

$$m_z = \frac{1}{2c} \sum_i e_i (x_i \dot{y}_i - y_i \dot{x}_i) \quad , \tag{1.4}$$

and the linearity in the velocities is preserved under any transformation to another set of generalized coordinates. The magnetic moment in the direction of the applied field (which defines the z-direction) is then given by

$$M_z = CN \int \cdots \int \sum_{k=1}^{f} a_k \dot{q}_k e^{-\mathcal{H}/k_B T} dq_1 ... dq_f dp_1 ... dp_f \quad . \tag{1.5}$$

Since Hamilton's equations relate the velocities to the momenta via

$$\dot{q}_f = \frac{\partial \mathcal{H}}{\partial p_j} \quad , \quad \dot{p}_f = -\frac{\partial \mathcal{H}}{\partial q_j} \quad , \tag{1.6}$$

the integrand for a particular index j is merely

$$-k_B T \frac{\partial \left(a_j e^{-\mathcal{H}/k_B T} \right)}{\partial p_j} \tag{1.7}$$

so that for this value j the integral in (1.5) becomes

$$-CN k_B T \int \cdots \int \left[a_j e^{-\mathcal{H}/k_B T} \right]_{p_j=-\infty}^{p_j=+\infty} dq_1 ... dq_f dp_1 ... dp_{j-1} dp_{j+1} ... dp_f. \tag{1.8}$$

Ergodicity requires that the integration has to be carried out over the whole range of the phase space so that the value for p_j varies over $\pm\infty$. From the convexity property of the free energy (see Sect. A.) one deduces that the energy has to become infinite for infinite values of the coordinate p_j which makes the integrand zero for any particular coordinate p_j. This proof holds for any Hamiltonian $\mathcal{H}$ (also with an applied field) since no properties of $\mathcal{H}$ were required.

This result is not only the formal proof that classical mechanics cannot account for magnetism but it also asks for an explanation of why one is able to obtain the Langevin results, (1.1) and (1.2), from the same classical mechanics. The answer is both simple and complicated at once: To derive Langevin's formula one has already assumed that finite and constant magnetic moments are present. In the sense of the proof given above, this means that one has restricted the integration to particular parts of the phase space, or in other words, some relevant parts of the energy remain finite (even constant) while the coordinates go to infinity. While classical mechanics cannot give any reasoning for such a restriction, quantum mechanics can, e.g. by requiring quantized values for the angular momentum or finite occupation numbers. In this sense van Leeuwen's theorem not only proves the inability of classical mechanics to explain magnetism, but also calls for a different mechanics on the microscopic scale which justifies the requirements for the Langevin formulation.

Langevin's result (1.1) can of course be derived from first principles quantum mechanics where it is known under the name Langevin–Debye formula. In the chapter about the Weiss molecular field model (Chap. 6), it will be shown how the Langevin–Debye result (6.13) can be obtained. A more profound derivation together with an extensive discussion is given in the book by John van Vleck [15].

The first successful quantum mechanical theory of magnetism goes back to N. Bohr who, around 1913, developed what is now known as the old quantum mechanics. (The "new quantum mechanics" – starting about 1921 – was later pioneered by Heisenberg, Schrödinger, Born, and Dirac.) With Bohr's astonishingly simple theorem that the angular momentum is given by multiples of Planck's constant, he immediately solved all problems which classical statistics imposed on the physics of microscopic particles. Among these problems of classical physics one usually only refers to the electrodynamically forbidden stationary electron orbits, but there existed other unanswered questions which rely on the application of Maxwell–Boltzmann statistics. One of these problems is the distribution of atomic (or molecular) sizes, since if Maxwell–Boltzmann statistics applies, the distribution of atomic radii in an ensemble of atoms must vary from very small ones (even zero) to very large ones. In particular the behavior for very small radii resembles the problem of the black-body radiation where classical physics requires that the energy density goes to infinity for zero wavelength (Rayleigh–Jeans law). Since quantum mechanics can thus explain why some variables remain constant, as discussed by van Leeuwen, it can also account for magnetism.

The microscopic theory of ferromagnetism starts with Weiss [16] who in 1907 postulated his famous molecular field model. Although formulated before the advent of quantum mechanics, he already assumed discrete energy levels associated with respective values for magnetic moments (angular momenta). To explain spontaneous alignment of the elementary magnetic moments, he postulated the existence of an internal (molecular) field which should represent the then unknown interaction between these moments. Later this interaction was found to be the exchange interaction which again is of entirely quantum-mechanical origin.

With the advancing spectroscopic techniques, the explanation of atomic spectra required the introduction of a new quantum number. Landé [17] proposed that he could explain his g-factor by assuming that the atom contained a mysterious *Atomrumpf* whose magnetic moment is exactly half the value found for ordinary angular momenta. The actual idea that the electron has an internal degree of freedom is due to Uhlenbeck and Goudsmit [18]. Only with the spin did it become possible to understand the anomalous Zeeman effect (i.e. the non-linear splitting of spectral lines in a weak applied field). Although the spin was originally described in analogy to the other angular momenta its actual origin was discovered by Dirac on formulating the relativistic quantum mechanics [19].

With the knowledge of the four principle quantum numbers (angular l, magnetic m_l, spin s, and spin–magnetic m_s) and the two different ways of coupling these angular momenta in the non–relativistic (ll–coupling) and in the relativistic case (jj–coupling) it became possible to explain the multiplet structure of atoms and molecules. Hund [20] combined these efforts in his three famous rules which describe a simple way to determine the term symbol and thus the multiplet structure of a given atom (see Sect. I.)

In 1927 Heitler and London [21] (Sect. 16.1) published a quantum mechanical calculation for the H_2 molecule. They showed that on obeying the exchange interaction by antisymmetrization of the molecular wavefunctions, the ground state of H_2 is either a singlet or a triplet state, depending on the sign of the exchange integral J. For negative J the spins of the two electrons are parallel (triplet state – "ferromagnetic alignmet"), for positive J the electrons in the ground state have antiparallel spin (singlet state – "anti-ferromagnetic alignment") which also can be regarded as a "non-magnetic" state. This result suddenly gave the "molecular field" of the Weiss model a direct quantum mechanical origin.

The first quantum mechanical formulation of the interaction between two spins leading to magnetic order is the phenomenological Heisenberg Hamiltonian [22] (Chap. 7) where spins located on different sites interact via an exchange parameter (often called the exchange integral). Heisenberg also succeeded in demonstrating that the Weiss molecular field is just the exchange interaction which tends to align spins in parallel if the exchange integral is negative. Within the Heisenberg model both ferromagnetic and anti-ferromagnetic order can be described and at finite temperature a new kind of elementary excitations of the whole spin system (magnons) are found.

The first microscopic account of anti-ferromagnetism was given by Neél [24] who generalized the Weiss model for the anti-ferromagnetic case by assuming two sublattices for the collinear up and down spins. In this model a spin on one sublattice interacts with spins on the same and on the other sublattice giving rise to a two component susceptibility divided into a parallel and a perpendicular part.

All the models mentioned so far deal with localized moments. The carrier of magnetism is a spin or an angular momentum and thus the magnetic moments measured are given by the expectation value of the respective quantum numbers. Examples for this type of localized magnetism are free atoms and molecules, or single ions in solids. Among the metallic solids localized moments are found for the f-electrons of the rare earths. The understanding of metallic magnetism (in particular of the transition metals) remained a challenge for a longer time. The measured magnetic moments were non-integer, so that a relation to angular- or spin-moments was not straight forward since these quantities appear to be quenched (see Sect. C.). The development of the electron band model [25] paved the way for an understanding of the magnetism of "itinerant" electrons. The first attempts to understand the metallic state were based on the free electron gas model and Pauli could show that the quenching of the spin can be explained if one assumes that all the electrons in open shells (conduction electrons) are at least partially free. The Pauli exclusion principle requires that two electrons have to be different in at least one quantum number. For free electrons the only relevant quantum numbers are the spin and the electron momentum $\mathbf{k}$. Since any state with quantum number $\mathbf{k}$ can accommodate two electrons, these electrons must have opposite spin so that all spins appear to be compensated. For the non-interacting (no exchange) gas of free electrons Pauli [26] derived the susceptibility to be

$$\chi_P = 2\mu_B^2 N(\varepsilon_F) \quad , \tag{1.9}$$

which relates the paramagnetism of the free electrons to the density of states at the Fermi energy $N(\varepsilon_F)$. Formally this relation resembles the result for the specific heat of the free electron gas which also depends on the density of states at the Fermi energy. These ideas about the metallic state described as an electron gas were further developed by Mott [27] and Slater [28, 29] who introduced more realistic models for the band structure and finally by Stoner [30] who succeeded in formulating a phenomenological "molecular field" model analogous to the Weiss model, but by replacing the discrete angular momentum levels by the electronic band structure. This model of itinerant electron magnetism accounts for non-integer magnetic moments in terms of a band filling of the narrow d-band and by taking into account the interaction of the s- and d-electrons of the valence band. Within the framework of the Stoner model (see Chap. 8), the susceptibility of the interacting free electron gas is enhanced by the exchange interaction and reads

$$\chi_S = \frac{2\mu_B^2 N(\varepsilon_F)}{1 - I_s N(\varepsilon_F)} \quad . \tag{1.10}$$

The quantity I_s in (1.10) is the Stoner exchange factor which is a slowly varying atomic quantity through the periodic table. The denominator allows one to formulate a criterion for the spontaneous onset of magnetism. If, in the paramagnetic state, the denominator is negative the resulting negative susceptibility means that the paramagnetic state is not at a total energy minimum but at a maximum, so that any magnetic state must have a lower total energy. This is the famous Stoner criterion which is usually formulated such that magnetism occurs if the following inequality holds

$$I_s N(\varepsilon_F) \geq 1 \quad . \tag{1.11}$$

By performing Hartree–Fock and tight binding calculations for the free electron gas attempts were made to clarify the role of the exchange interaction in metals (see e.g. the review by Wohlfarth [31] and references given therein). A crucial step towards understanding and towards the practicability of quantum mechanical calculations of the metallic state was made by Slater [32] who suggested that the exchange interaction should be approximated by an averaged potential over the occupied states of a homogenous electron gas which is no longer a non–local quantity but depends only on the local electron density at the point $\mathbf{r}$. The exchange potential energy then reads

$$v_S(\boldsymbol{r}) = -3e^2 \left(\frac{3}{8\pi}\right)^{\frac{1}{3}} \rho(\boldsymbol{r})^{\frac{1}{3}} \quad . \tag{1.12}$$

A significant improvement to the Slater exchange occurred with the introduction of the X_α method where the Slater exchange potential $v_S(\boldsymbol{r})$ is multiplied by a factor which is determined by the requirement that the total energy of an isolated atom calculated from the X_α–potential equals the respective Hartree–Fock value [33].

The next major step towards an understanding of the electronic structure of solids and subsequently of their magnetism was the density functional formalism introduced by Hohenberg and Kohn [34]. They showed that ground state energy can be expressed in terms of an universal functional of the electron density. In a second paper it was shown that the variation of the energy with respect to the electron density leads to effective one-electron Schrödinger type equations [35]. In these equations the exchange-correlation potential $v_{xc}(\boldsymbol{r})$ enters additively to the coulomb part, which makes it straight forward to use independent interpolation formulae to approximate $v_{xc}(\boldsymbol{r})$. This treatment is called the local-spin-density-approximation (LSDA) for exchange and correlation [36, 37, 38].

Within the framework of the local-density-approximation the Stoner exchange factor I_s can be evaluated in a straight forward way. It appears that the exchange splitting ΔE can be written as the expectation value for the difference of the spin-up and spin-down exchange–correlation potential

$$I_s M = \Delta E = \langle \Psi \mid v_{xc}^\uparrow - v_{xc}^\downarrow \mid \Psi \rangle. \tag{1.13}$$

where I_s is the Stoner exchange factor and M is the resulting magnetic moment. $v_{xc}^\uparrow$ and $v_{xc}^\downarrow$ are the exchange–correlation potentials for spin-up and spin-down respectively. In Table 1.1 the values for ΔE and I_s for hcp- (fcc-) cobalt for some commonly used models of the local-spin-density-approximation are compared. (a) best value from experiments (hcp Co) given by Wohlfarth [39]; (b) X_α result for fcc Co by Wakoh and Yamashita [40]; (c) to obtain an improved description of the correlation effects Oles et al. [41] introduced their local approach (LA) (fcc Co); (d),(e) values taken from own calculations for hcp Co employing the Hedin–Lundquist (HL) [37] and

Table 1.1. Exchange splitting ΔE and Stoner factor I_s for closed packed cobalt for various models of the local density approximation for exchange and correlation. Despite of the large scattering found for ΔE and I_s the calculated magnetic moments are all between 1.55 and $1.7\mu_B$ (exp: $1.62\mu_B$). For the meaning of the superscripts see preceding text

	expt.[a]	$X\alpha^{[b]}$	LA[c]	HL[d]	vBH[e]
ΔE(eV)	1.05	2.05	1.23	1.40	1.48
I_s(eV/μ_B)	0.65	1.21	0.72	0.93	0.96

the von Barth–Hedin (vBH) [38] exchange potential respectively. It appears that most of the usual models yield too large a band splitting and Stoner factor. This shortcoming is due to an incomplete description of the correlation effects. Only in case (**c**) where a special treatment is used to improve the correlation do the calculated values come closer to the experimental estimates.

Parallel to the development of band structure theory there was a search for "simple" toy-models to reproduce solid state magnetism. These models should reproduce the main features of electrons in a solid which are the bonding to a certain site and the coulomb repulsion of electrons at the same site. The most prominent of these models was introduced by Hubbard [42] (Chap. 13) who tried to describe transition metal oxides with their narrow d-bands and their strong correlation.

With these tools at hand it became possible to calculate the ground state properties with high reliability and also to understand the complex mechanisms occurring in solids. Among the remaining problems the most crucial one was the temperature dependence of solid state magnetism. While for the localized moment models the finite temperature effects became reasonably clear at a very early stage, this development took some time for the delocalized electrons in most solids.

For the free electron gas the temperature dependence of the susceptibility was calculated from the finite temperature properties of the Fermi–Dirac distribution (Sommerfeld expansion). Stoner applied the same ideas to introduce finite temperatures into the itinerant electron model. Unfortunately the respective Curie temperatures came out too large by a factor of $4 - 8$ and the inverse susceptibility above the Curie temperature showed a T^2 rather than a linear T dependence (as observed experimentally). After 30 years of persistent struggle to salvage the Stoner model it finally became clear that the single particle excitations are not (or only to a small degree) responsible for the finite temperature behavior of metallic magnetism.

These results suggest that one must consider two extreme limits:

- the localized limit for which the magnetic moments and their fluctuations are localized in real space (delocalized in reciprocal space), with their amplitudes being large and fixed.
- the itinerant limit for which the moments and their fluctuations are localized in reciprocal space (delocalized in real space), with their amplitudes being temperature dependent.

A Curie–Weiss law is observed in both cases but its physical origin and the corresponding value of the Curie constant will be different.

To solve these inconsistencies Moriya tried to include thermally induced collective excitations of the spin systems (as they were already known for localized spins) in order to formulate an unified picture of magnetism [43]. A similar very promising approach was introduced by Murata and Doniach [44] who introduced local and random classical fluctuations of the spin-density (spin fluctuations) which should be excited thermally. The latter two models become equivalent at high temperatures and lead to a Curie–Weiss law. Although the Murata–Doniach approach runs into trouble at $T = 0K$ (the classical fluctuations cause a violation of the 3rd law of thermodynamics) it has successfully been applied to calculate Curie temperatures from data derived from de Haas–van Alphen measurements [46] and to formulate a simple model for the Curie temperature of metallic solids [47] which gives much better results than the above mentioned Stoner model.

Today, about a hundred years after Langevin, there exists a fairly good knowledge about the basic mechanisms of localized and itinerant electron magnetism but many open questions still remain. A practicable unified picture of magnetism is still not at hand. The magnetism of strongly correlated systems is not very well understood. New effects (e.g. the high temperature superconductors) create new questions and new technologies create new materials like the magnetic superstructures with very interesting new properties such as giant magnetoresistance. As in all disciplines magnetism is no longer a subject *sui generis*. It is entangled with electronic structure and correlation effects, with crystalline phases and their stability and with dynamical processes like the excitation of collective modes and relaxation effects on long time scales. All this adds up to make magnetism one of the most interesting and most challenging subjects of solid state physics.

2. Consequences of Fermi Statistics

2.1 Quantum Statistics of Fermions

To describe the quantum mechanical properties of an ensemble of particles one has to choose the appropriate statistics. The kind of statistics is of course a consequence of the assumptions one makes. Fermi statistics is based on three axioms:

- The particles are indistinguishable
- The particles should obey the Pauli principle (exchange interaction); one thus deals with particles whose spin is given by $n + 1/2$ with $n \in N$; the particle wavefunctions are antisymmetric against the exchange of two particles.
- Interactions between the particles are only weak. This assumption may cause trouble (see e.g. the ideal gas). To describe collective excitations (magnons) it becomes necessary to change the statistics (Bose statistics).

At an average energy ε_s, a system has a number of quantum states a_s with occupation numbers n_s. The number of different distributions is given by

$$W_s = \frac{a_s!}{n_s!\,(a_s - n_s)!} \quad . \tag{2.1}$$

The division by $n_s!$ accounts for the fact that the particles are indistinguishable. (e.g. $a_s = 3$, $n_s = 2, \Rightarrow W_s = 3$. If one would have used classical Maxwell–Boltzmann statistics $W_s = a_s^{n_s} = 9$.) W_s is of course proportional to the probability of a state being occupied or not and can replace this probability, because for the maximization of the entropy only the logarithmic derivative is needed. For the total probability W, the number of particle states N and the total energy E one finds:

$$W = \prod_s W_s = \prod_s \frac{a_s!}{n_s!\,(a_s - n_s)!} \quad , \tag{2.2}$$

$$N = \sum_s n_s \quad , \tag{2.3}$$

$$E = \sum_s \varepsilon_s n_s \quad . \tag{2.4}$$

One now has to maximize the entropy (maximize W) under the condition that the number of particles N, and for a closed system also the total energy is constant. From Boltzmann's formula $S = k_B \ln(W)$ one obtains the variation

$$dS = k_B d(\ln(W)) = k_B \left(\frac{\partial \ln(W)}{\partial n} \right) dn \quad ,$$

where only the variation with respect to the occupation number n needs to be considered, because all other variables can be expressed in terms of n. It is thus only necessary to calculate the quantity $\partial \ln(W)/\partial n$, because W is the probability for the "most probable" configuration of the system. The variation is written as

$$\partial \ln(W) - \alpha \partial N - \beta \partial E = 0 \quad , \tag{2.5}$$

introducing the Lagrange multipliers α and β. Employing the relations (2.2)–(2.4) one obtains

$$\partial N = \sum_s \partial n_s \quad ,$$

$$\partial E = \sum_s \varepsilon_s \partial n_s \quad . \tag{2.6}$$

The calculation of $\ln W$ is more elaborate:

$$\ln W = \sum_s \ln W_s = \sum_s \ln \frac{a_s!}{n_s! \, (a_s - n_s)!} \quad .$$

Using Stirling's approximation for the logarithm of the factorial

$$\ln(k!) \simeq k \ln(k) - k \quad , \tag{2.7}$$

one gets

$$\ln W = \sum_s \ln(a_s!) - \ln(n_s!) - \ln(a_s! - n_s!)$$

$$= \sum_s n_s \ln \left(\frac{a_s}{n_s} - 1 \right) - a_s \ln \left(1 - \frac{n_s}{a_s} \right) \quad . \tag{2.8}$$

One thus obtains for the variational condition (2.5)

$$\sum_s \left(\ln \left(\frac{a_s}{n_s} - 1 \right) - \alpha - \beta \varepsilon_s \right) \partial n_s = 0 \quad . \tag{2.9}$$

Because of the principle of detailed equilibrium the variational condition has to be fulfilled for any value of s giving:

$$\ln \left(\frac{a_s}{n_s} - 1 \right) - \alpha - \beta \varepsilon_s = 0 \quad , \tag{2.10}$$

and hence

$$n_s = \frac{a_s}{\exp(\alpha + \beta\varepsilon_s) + 1} = a_s f(\varepsilon_s) \quad . \tag{2.11}$$

The assumptions for the quantum statistics thus lead to the Fermi–Dirac distribution. If one would have used classical statistics for the probability (2.1) $W = a_s^{n_s}$, one would have obtained the Boltzmann distribution

$$n_s = \exp(-\alpha - \beta\varepsilon_s) \quad . \tag{2.12}$$

To obtain the sum of states one uses (2.11) and (2.8)

$$\ln W = \sum_s [n_s(\alpha + \beta\varepsilon_s) + a_s \ln(1 + \exp(-\alpha - \beta\varepsilon_s)] \quad . \tag{2.13}$$

Taking the total derivative of the energy E

$$\partial E = \sum_s \varepsilon_s \partial n_s + \sum_s n_s \partial \varepsilon_s \quad , \tag{2.14}$$

one can use (2.14) to introduce the thermodynamical variables entropy TdS and work dW' which are used to identify the first an second part of the energy variation respectively. Using this identification one can rewrite $\ln W$ as:

$$\partial \ln W = \sum_s \alpha \partial n_s + \sum_s \beta \varepsilon_s \partial n_s \quad . \tag{2.15}$$

Under the assumption that the number of particles is constant one can put $\alpha \sum_s \partial n_s = 0$. This does not hold for the second term, because there one has to sum over the product of ∂n_s times the "spectral weight" energy. Putting everything together yields a differential form of Boltzmann's formula

$$\partial \ln W = \beta T \partial S \quad . \tag{2.16}$$

Equation (2.16) also implicitly defines the Lagrangian multiplier β to take the value $1/(k_B T)$. The entropy can now be written as

$$\frac{S}{k_B} = \ln W = \sum_s [n_s(\alpha + \beta\varepsilon_s) + a_s \ln(1 + \exp(-\alpha - \beta\varepsilon_s)] \quad . \tag{2.17}$$

Carrying out the summation one obtains

$$\frac{S}{k_B} = \alpha N + \beta E + Z \quad ,$$

$$Z = \sum_s a_s [\ln(1 + \exp(-\alpha - \beta\varepsilon_s))] \quad . \tag{2.18}$$

The quantity Z is called the "sum of states" or "partition function". Z is a central quantity of statistical thermodynamics; some of its properties, which will be used later on, are given in brief:

$$N = \sum_s n_s = -\frac{\partial Z}{\partial \alpha} \quad,$$

$$P = k_B T \frac{\partial Z}{\partial V} \quad, \tag{2.19}$$

$$T\partial S = \partial E + k_B T \partial Z = \partial E + P \partial V \quad.$$

2.2 Free Energy of the Fermi Gas

From standard thermodynamics one uses the relations for the free energy F

$$F = E - TS = -k_B T (N\alpha + Z) \quad, \tag{2.20}$$

and the free enthalpy G

$$G = F + PV = -k_B T (N\alpha + Z) + k_B T Z$$
$$= -Nk_B T\alpha \quad. \tag{2.21}$$

Equation (2.21) provides an immediate definition for the Lagrangian multiplier α which can be written in terms of the chemical potential μ,

$$\frac{G}{N} = -\alpha k_B T = \mu \quad.$$

Originally a_s denoted the number of quantum states at an energy ε_s. If one now introduces quasi-continuous energies, one replaces a_s by the density of states (DOS) $\mathcal{N}(\varepsilon)$. Since $\mathcal{N}(\varepsilon)$ counts the number of states in the energy interval $d\varepsilon$ one can write

$$a_s \Rightarrow \mathcal{N}(\varepsilon)\, d\varepsilon \quad,$$
$$n_s \Rightarrow \mathcal{N}(\varepsilon) f(\varepsilon)\, d\varepsilon \quad.$$

In solid state physics the density of states is a central quantity for the understanding of the electronic and magnetic properties. The density of states is the solid state analogue of the energy level in a free atom with a certain occupation number. The density of states is calculated via the Brillouin zone integration over the band structure of a solid (i.e. the dispersion relation between the electron momentum $\boldsymbol{k}$ and the respective energy eigenvalues), as shown in Chap. 4.

For $T \neq 0$ the DOS becomes multiplied with the Fermi–Dirac distribution and reads

$$\mathcal{N}(\varepsilon) f(\varepsilon) = \mathcal{N}(\varepsilon) \frac{1}{\exp\left(\frac{\varepsilon-\mu}{k_B T}\right) + 1} \quad. \tag{2.22}$$

With the density of states $\mathcal{N}(\varepsilon)$ the sum of states becomes

$$Z = \int_0^\infty \mathcal{N}(\varepsilon) \ln\left(1 + \exp\left(\frac{\mu - \varepsilon}{k_B T}\right)\right) d\varepsilon \quad . \tag{2.23}$$

For the gas of free electrons, the density of states is of the form

$$\mathcal{N}(\varepsilon) = \frac{4\pi}{h^3} (2m)^{\frac{3}{2}} \varepsilon^{\frac{1}{2}} \quad . \tag{2.24}$$

In (2.24) it was not distinguished between the two spin directions. At $T = 0$K the electron gas is of course in the state of lowest energy. All states below the Fermi energy ε_F are occupied, all states above are empty. This is caused by the form of the Fermi distribution for $T = 0$K which is 1 below ε_F and 0 above. At $T = 0$K, ε_F is equal to the chemical potential μ which is defined as the energy necessary to add one particle to the system. As all states below ε_F are occupied the lowest unoccupied state (in a metal) has exactly the energy ε_F. This definition has to be changed if one considers semiconductors or insulators where the lowest unoccupied state is separated from the bottom of the conduction band by the insulating (or semiconducting) gap. In semiconductor physics it is common to place the Fermi energy in the middle of the gap, which, however, no longer allows an interpretation of ε_F as the chemical potential.

The concept of the Fermi energy and the analysis of the shape of the density of states at ε_F enter into a number of properties of the solid. The reason for the major role of the Fermi surface is again the fact that it is the threshold between occupied and unoccupied states, the only place in phase space where excitations (thermal, magnetic, etc.) can occur.

The free electron DOS (Fig. 2.1) has the shape of a parabola. Very often the expression parabolic band is used as a synonym for a free electron like behavior if the energy is proportional to k^2 (square of the electron momentum). As an example the DOS of fcc-aluminium is shown in Fig. 2.2. The nearly parabolic band is obvious and is typical for valence electrons (s, p electrons) of simple metals (without d-electrons) which show a free electron like

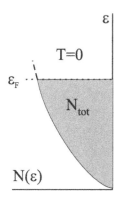

Fig. 2.1. Parabolic density of states (DOS) of the free electron gas at $T = 0$

behavior. The "ripples" around the Fermi energy are caused by the actual electronic bandstructure and are called "van Hove–singularities". This free electron model for the simple metals enabled physicists to describe their basic properties long before band structure calculations were available.

The number of particles (electrons) up to the energy ε_F is given by

$$N_{\text{tot}} = \int_0^{\varepsilon_F} \mathcal{N}(\varepsilon)\mathrm{d}\varepsilon \quad . \tag{2.25}$$

Conversely if the number of particles is known, the Fermi energy ε_F can always be calculated from the DOS. Again for the free electron gas one obtains

$$\varepsilon_F = \frac{h^2}{2m}\left(\frac{3N_{\text{tot}}}{8\pi V}\right)^{\frac{2}{3}} \quad . \tag{2.26}$$

One uses ε_F to define a temperature $T_F = \varepsilon_F/k_B$ which is called the Fermi degeneracy temperature. T_F sets the scale for the energies in the solid and defines a relation between the temperature of the system and the temperature dependence of the various physical properties. T_F is of the order of $10^4 - 10^5$K for sodium and other simple and transition metals and only for a few rare earth systems (mainly Ce-containing *heavy fermion* systems) T_F goes down to 10^3K. Why is this characteristic temperature so important? Since the Fermi energy ε_F is equal to the chemical potential μ (at $T = 0$) T_F relates the energy needed to add one particle to the system to excitation energies. Thermal excitations are usually of the order of room temperature, excitations due to magnetic or electric fields are at least one order of magnitude smaller (expressed in terms of energy, an external magnetic field of 1T corresponds to a temperature of about 0.6 K). These relations immediately show that the electronic structure of a solid usually remains fairly unaffected by an excitation of normal magnitude. This property of fermionic systems will be used again when one describes the response of a solid to excitations solely from the properties of the DOS around the Fermi energy.

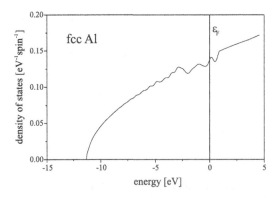

Fig. 2.2. Density of states of fcc aluminium

At finite temperature however, the DOS becomes modified by the Fermi–Dirac distribution (Fig. 2.3)

$$N(\varepsilon, T) = \frac{3}{2} \frac{N_{\text{tot}}}{\varepsilon_{\text{F}}^{\frac{3}{2}}} \frac{\varepsilon^{\frac{1}{2}}}{\left(\exp\left(\frac{\varepsilon - \mu}{k_{\text{B}}T} \right) + 1 \right)} \quad , \quad \mu = \mu(T) \quad . \tag{2.27}$$

It should be noticed that Fig. 2.3 strongly exaggerates the influence of temperature on the Fermi surface. The change due to the finite temperature Fermi–Dirac distribution occurs in an energy interval of $k_{\text{B}}T$ around the Fermi energy. A distribution as shown in Fig. 2.3 would thus be due to a temperature of several thousand K.

To derive an expression for the electron gas at finite temperatures, one introduces the abbreviations

$$x = \frac{\varepsilon}{k_{\text{B}}T} \quad , \quad \eta = \frac{\mu}{k_{\text{B}}T} \quad , \quad n = \frac{N_{\text{tot}}}{V} \quad . \tag{2.28}$$

For the number of particles n one obtains

$$N_{\text{tot}} = \int_0^\infty N(\varepsilon, T) d\varepsilon$$

$$= \frac{3}{2} N_{\text{tot}} \left(\frac{k_{\text{B}}T}{\varepsilon_{\text{F}}} \right)^{\frac{3}{2}} F_{1/2}(\eta) \quad , \tag{2.29}$$

where $F_{1/2}(\eta)$ is the Fermi integral which in general is defined as

$$F_y(\eta) = \int_0^\infty \frac{x^y}{\exp(x - \eta) + 1} dx \quad . \tag{2.30}$$

In the latter two equations one has to replace the upper limit of integration by infinity since for $T > 0$ a Fermi energy is no longer rigorously defined. Equation (2.29) allows one to calculate the value of $F_{1/2}(\eta)$

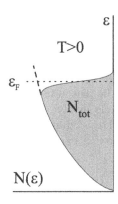

$N(\varepsilon)$

Fig. 2.3. Free electron density of states at finite temperature

$$1 = \frac{3}{2}\left(\frac{k_B T}{\varepsilon_F}\right)^{\frac{3}{2}} F_{1/2}(\eta) \quad , \quad \eta = \eta(T) = \frac{\mu}{k_B T} \quad ,$$

$$F_{1/2}(\eta(T)) = \frac{2}{3}\left(\frac{\varepsilon_F}{k_B T}\right)^{\frac{3}{2}} \quad . \tag{2.31}$$

In an analogous way one can calculate the energy of the occupied states giving

$$E = \int_0^\infty \varepsilon \mathcal{N}(\varepsilon, T)\mathrm{d}\varepsilon = \frac{3}{2} N_{tot} k_B T \left(\frac{k_B T}{\varepsilon_F}\right)^{\frac{3}{2}} F_{3/2}(\eta) \quad , \tag{2.32}$$

$$\Rightarrow \frac{E}{N_{tot} k_B T} = \frac{F_{3/2}(\eta)}{F_{1/2}(\eta)} \quad , \tag{2.33}$$

with the sum of states

$$Z = \frac{3}{2}\left(\frac{k_B T}{\varepsilon_F}\right)^{\frac{3}{2}} N_{tot} \int_0^\infty x^{\frac{1}{2}} \ln\left(1 + \exp\left(\eta - x\right)\right) \mathrm{d}x$$

$$= N_{tot}\left(\frac{k_B T}{\varepsilon_F}\right)^{\frac{3}{2}} F_{3/2}(\eta) \tag{2.34}$$

$$= \frac{2}{3} N_{tot} \frac{F_{3/2}(\eta)}{F_{1/2}(\eta)} = \frac{2}{3}\frac{E}{k_B T} \quad .$$

Now one expresses the thermodynamical variables entropy S, free energy F, and enthalpy G in terms of the Fermi integrals

$$\frac{S}{N_{tot} k_B} = -\eta + \frac{5}{3}\frac{F_{3/2}(\eta)}{F_{1/2}(\eta)} \quad , \tag{2.35}$$

$$\frac{F}{N_{tot} k_B T} = \eta - \frac{2}{3}\frac{F_{3/2}(\eta)}{F_{1/2}(\eta)} \quad , \tag{2.36}$$

$$\frac{G}{N_{tot} k_B T} = -\alpha = \frac{\mu}{k_B T} = \eta(T) \quad . \tag{2.37}$$

To obtain the formulae given above, a free electron density of states was assumed. However, one finds that these relations also hold quite well for general densities of states as e.g. in the transition metals. The reason for this behavior is that, because of the Fermi distribution, all effects involve states in a small energy range around ε_F, because T_F is much larger than room temperature. In this case $\eta \gg 1$ and thus $\frac{\mu}{k_B T} \gg 1$ one can expand the Fermi functions as

$$F_y(\eta) = \frac{\eta^{y+1}}{y+1}\left[1 + 2\sum_{r=1}^{\infty}\frac{(y+1)\,y...\,(y-2r+2)}{\eta^{2r}}\left(1 - 2^{1-2r}\zeta(2r)\right)\right],$$

$$(2.38)$$

where $\zeta(2r)$ is the Riemann ζ-function.

For the Fermi function one obtains in lowest order in T (expressed in terms of η) the approximations

$$F_{1/2}(\eta) \cong \frac{2}{3}\eta^{\frac{3}{2}}\left(1 + \frac{\pi^2}{8}\eta^{-2} + ...\right),$$

$$F_{3/2}(\eta) \cong \frac{2}{5}\eta^{\frac{5}{2}}\left(1 + \frac{5\pi^2}{8}\eta^{-2} + ...\right).$$

$$(2.39)$$

Employing (2.29) one can now derive an expression for the temperature dependence of the chemical potential

$$F_{1/2}(\eta(T)) = \frac{2}{3}\left(\frac{\varepsilon_F}{k_BT}\right)^{\frac{3}{2}} = \frac{2}{3}\eta^{\frac{3}{2}}\left(1 + \frac{\pi^2}{8}\eta^{-2}\right).$$

Since

$$\frac{\varepsilon_F}{k_BT} = \frac{\mu(0)}{k_BT} = \eta, \qquad (2.40)$$

one obtains

$$\eta = \frac{\mu(T)}{k_BT} = \frac{\mu(0)}{k_BT}\left(1 - \frac{\pi^2}{8}\left(\frac{k_BT}{\varepsilon_F}\right)^2\right)^{\frac{2}{3}}. \qquad (2.41)$$

Expanding also $(1-x)^{2/3}$ one yields a physically intuitive approximation of the form

$$\mu(T) \simeq \mu(0)\left(1 - \frac{\pi^2}{12}\left(\frac{k_BT}{\varepsilon_F}\right)^2\right). \qquad (2.42)$$

At elevated temperature (2.42) describes a lowering of the chemical potential. This behavior is due to the thermal excitation of particles into higher unoccupied states, making unoccupied states available at energies below ε_F which can be filled at an energy below $\mu(0)$. Since this effect again scales with T_F it can be expected to be rather small but it is nevertheless responsible for the specific heat of the electron gas. From (2.33), (2.39), and (2.42) one calculates an expression for the total energy E and the specific heat at constant volume, c_v

$$\frac{E}{N_{tot}\varepsilon_F} = \frac{3}{2}\left(\frac{k_BT}{\varepsilon_F}\right)^{\frac{5}{2}}F_{1/2}(\eta) = \frac{3}{5}\left(1 + \frac{5\pi^2}{12}\left(\frac{k_BT}{\varepsilon_F}\right)^2\right), \qquad (2.43)$$

$$c_v = \left(\frac{\partial E}{\partial T}\right)_V = N_{tot}\varepsilon_F\frac{3}{5}\frac{5\pi^2}{12}2\frac{k_B^2}{\varepsilon_F^2}T. \qquad (2.44)$$

If one writes c_v in the usual form as $c_v = \gamma T$, one finds for the coefficient of the linear term of the electronic specific heat

$$\gamma = \frac{\pi^2}{2} \frac{n k_B}{T_F} \quad , \quad T_F = \frac{\varepsilon_F}{k_B} \quad . \tag{2.45}$$

By means of (2.24) and (2.26) one expresses the factor γ of the electronic part of the specific heat terms of the density of states at the Fermi energy

$$(2.24) \rightarrow \mathcal{N}(\varepsilon) = \frac{4\pi}{h^3} (2m)^{\frac{3}{2}} \varepsilon^{\frac{1}{2}} \tag{2.46}$$

$$(2.26) \rightarrow \varepsilon_F = \frac{h^2}{2m} \left(\frac{3 N_{tot}}{8\pi V} \right)^{\frac{2}{3}}$$

$$\Rightarrow \mathcal{N}(\varepsilon_F) = \frac{3}{2} \frac{n}{\varepsilon_F} \quad ,$$

$$\gamma = \frac{\pi^2}{3} k_B^2 \mathcal{N}(\varepsilon_F) \quad . \tag{2.47}$$

The specific heat contribution of the valence electrons is proportional to the density of states at ε_F. Although our derivation was based on the free electron gas one finds that (2.47) also holds quite well for d-electron systems. The only necessary condition is that the valence electrons should show a quasi free behavior (itinerant electrons). In systems where these electrons form localized states and/or show strong correlation effects, (2.47) is no longer valid. Such a counter example are the heavy fermion systems, which due to the large effective mass of the electrons (fermions) at ε_F have an extremely large specific heat, which cannot be explained solely by the DOS at ε_F.

As an example for which (2.47) should be valid, we consider the specific heat coefficient of the simple metal Al. The calculated value is: $\gamma = 2.38 \times 10^{-4}$cal/mol/K^2 whereas the observed one amounts to: $\gamma = 4.18 \times 10^{-4}$ cal/mol/K^2. The reason for the discrepancy between the values lies in the electron–electron and the electron–phonon interaction which also gives a linear contribution to c_v. In a very crude way one can include these contributions as

$$\gamma_{obs} = \gamma_{calc} (1 + \lambda) \quad , \quad 0 \leq \lambda \leq 1 \quad . \tag{2.48}$$

How does the free electron gas behave at elevated temperatures? From the derivation above one finds that the specific heat rises linearly with temperature. For higher temperatures one has to consider higher order terms in the expansion of the Fermi integral and finds

$$\gamma = \frac{\pi^2}{3} k_B^2 \mathcal{N}(\varepsilon_F) \left(1 + b T^2 \right) \quad ,$$

$$b = \frac{\pi^2 k_B^2}{10} \left[7 \frac{\mathcal{N}(\varepsilon_F)''}{\mathcal{N}(\varepsilon_F)} - 5 \left(\frac{\mathcal{N}(\varepsilon_F)'}{\mathcal{N}(\varepsilon_F)} \right)^2 \right] \quad . \tag{2.49}$$

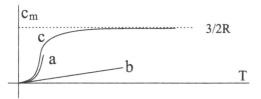

Fig. 2.4. Specific heat of the electron gas: (a) T^3 contribution, (b) linear contribution, (c) overall behavior

In the constant b appear the first and second derivatives of the density of states at the Fermi energy. This result shows that even at high temperatures the specific heat depends on the ground state properties at $T = 0K$ and is given by the particular details of $\mathcal{N}(\varepsilon_F)$. The expansion given in (2.49) is often called the Sommerfeld expansion. The characteristic form of (2.49) will appear again when one calculates the temperature dependence of the susceptibility. For the experimental evaluation of the T^3 term there is however the problem that there is usually a T^3 contribution from the phonons (Debye term) as well, which makes it hard to distinguish between the electronic and the phononic part. Figure 2.4 shows this behavior.

In the high temperature limit when T is large compared to T_F so that $\eta \ll 0$ the Fermi integrals are expanded asymptotically

$$F_y(\eta) = \Gamma(y+1) \sum_{r=1}^{\infty} (-1)^{r-1} \frac{\exp(r\eta)}{r^{y+1}} \quad , \tag{2.50}$$

so that for $\eta \to \infty \Rightarrow F_y(\eta) \to 0$. From (2.33) one obtains

$$\frac{E}{nk_BT} = \frac{\Gamma\left(\frac{5}{2}\right)}{\Gamma\left(\frac{3}{2}\right)} = \frac{3}{2} \quad ,$$

$$\Rightarrow E = \frac{3}{2}RT \quad , \quad c_v = \frac{3}{2}R \quad . \tag{2.51}$$

In the high temperature limit the specific heat contribution of the electrons becomes constant and approaches the classical value of the Dulong–Petit law as shown in Fig. 2.4. From the previous discussion it is obvious that this high temperature limit is experimentally never reached, because it requires that the temperature of the electron gas must be larger than T_F. For all practical applications the assumption of a linear contribution to the specific heat strictly holds until the melting temperature.

3. Paramagnetism

This chapter will describe how the free electron gas reacts to an external magnetic field. The model used is based on a classical picture where it is assumed that the magnetic moment caused by the spin can be described as a tiny "elemental magnet" (which is not even half the truth). In this case it will be energetically favorable for the spin to orient itself parallel to an applied field. Opposing this orientation is the loss in kinetic energy due to the occupation of states at higher energy.

At this point it is usually helpful to say a few words on the term "spin direction". In a non-relativistic treatment the direction of the spin is not given by a direction in the $\mathcal{R}3$. A change of the spin direction can thus never be obtained by an operation within the $\mathcal{R}3$. That means spin up in Austria is also spin up in Australia and vice versa. However if spin and angular momentum are coupled via the relativistic $\boldsymbol{LS}$ coupling, the direction of the spin is also coupled to the crystal axis. This in turn leads to an anisotropy of the magnetization which is manifested by the existence of a spatial dependence of the susceptibility. Experimentally this behavior is manifested by a preferred direction of the magnetic moment in a sample, which is the reason for the properties of all permanent magnet materials.

In equilibrium and without an applied magnetic field, each electronic state is occupied by two electrons with opposite spin. In a Gedankenexperiment this degeneracy is lifted and one observes the two spin directions separately

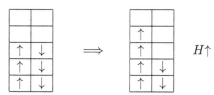

To include the magnetization and the other magnetic quantities in the equations derived in the previous chapter, one needs a few alterations. Including a magnetic field H in the free energy is accomplished by a term $-MH$, where M is the magnetic moment of the system caused by the field H. From this it follows that

$$M = -\left(\frac{\partial F}{\partial H}\right)_{T,V} \quad , \tag{3.1}$$

Using (2.20) one obtains easily

$$F = nk_BT\eta - k_BTZ \Rightarrow M = k_BT\frac{\partial Z}{\partial H} \quad . \tag{3.2}$$

The relation given in (2.23) has to be modified to contain the influence of the magnetic field H and to hold for both spin directions. As the density of states $\mathcal{N}(\varepsilon)$ is now split into two spin dependent contributions, the sum of states has to be carried out over spin up and spin down separately

$$Z = \int_0^\infty \mathcal{N}(\varepsilon)\ln\left(1 + \exp\left(\frac{\mu - \varepsilon - \mu_BH}{k_BT}\right)\right)d\varepsilon \tag{3.3}$$

$$+ \int_0^\infty \mathcal{N}(\varepsilon)\ln\left(1 + \exp\left(\frac{\mu - \varepsilon + \mu_BH}{k_BT}\right)\right)d\varepsilon \quad .$$

Since one accounts for each spin separately, the density of states $\mathcal{N}(\varepsilon)$ is thus divided by 2 and reads $\mathcal{N}(\varepsilon) = \frac{3}{4}n\varepsilon^{\frac{1}{2}}/\varepsilon_F^{3/2}$. In (3.3) the Bohr-magneton μ_B was introduced which is defined as

$$1\mu_B = \frac{e\hbar}{2m_e} = 0.578 \times 10^{-4}\mathrm{eV/T} = 0.927 \times 10^{-23}\mathrm{J/T} \quad , \tag{3.4}$$

which relates the magnetic field H to the energy.

One now carries out the same derivation for the gas of free electrons as in Sect. 2.2. To this end one introduces the following abbreviations

$$x = \frac{\varepsilon}{k_BT} \quad , \quad \eta = \frac{\mu}{k_BT} \quad , \quad \beta = \frac{\mu_BH}{k_BT} \quad , \tag{3.5}$$

and obtains

$$Z = \frac{1}{2}n\left(\frac{k_BT}{\varepsilon_F}\right)^{\frac{3}{2}}\left(F_{3/2}(\eta + \beta) + F_{3/2}(\eta - \beta)\right) \quad . \tag{3.6}$$

To calculate the derivative of Z the recursion relation for the derivatives of the Fermi integrals is applied

$$\frac{dF_k(\eta)}{d\eta} = kF_{k-1}(\eta) \quad . \tag{3.7}$$

Using the relations $M = \mu_B\frac{\partial Z}{\partial \beta}$ and $n = \frac{\partial Z}{\partial \eta}$ in connection with (3.7) one obtains

$$M = \frac{3}{4}n\mu_B\left(\frac{k_BT}{\varepsilon_F}\right)^{\frac{3}{2}}\left(F_{1/2}(\eta + \beta) - F_{1/2}(\eta - \beta)\right) \quad ,$$

$$n = \frac{3}{4}n\left(\frac{k_BT}{\varepsilon_F}\right)^{\frac{3}{2}}\left(F_{1/2}(\eta + \beta) + F_{1/2}(\eta - \beta)\right) \quad . \tag{3.8}$$

One now introduces the relative magnetization ζ as the ratio between the magnetic moment and the total number of particles

$$\zeta = \frac{M}{n\mu_{\mathrm{B}}} = \frac{M}{M_0} \quad , \tag{3.9}$$

which can be expressed in terms of the Fermi integrals

$$\zeta = \frac{F_{1/2}\left(\eta+\beta\right) - F_{1/2}\left(\eta-\beta\right)}{F_{1/2}\left(\eta+\beta\right) + F_{1/2}\left(\eta-\beta\right)} = \frac{n^+ - n^-}{n^+ + n^-} \quad . \tag{3.10}$$

For $\beta = \mu_{\mathrm{B}} H/\left(k_{\mathrm{B}} T\right) << 1$ the Fermi integral is expanded in a Taylor series

$$F_{1/2}\left(\eta+\beta\right) \simeq F_{1/2}\left(\eta\right) + \beta\frac{\mathrm{d}F_{1/2}\left(\eta\right)}{\mathrm{d}\eta} = F_{1/2}\left(\eta\right) + \beta F_{1/2}'\left(\eta\right) \ ,$$

so that ζ becomes

$$\zeta = \frac{F_{1/2}\left(\eta\right) + \beta F_{1/2}'\left(\eta\right) - F_{1/2}\left(\eta\right) + \beta F_{1/2}'\left(\eta\right)}{F_{1/2}\left(\eta\right) + \beta F_{1/2}'\left(\eta\right) + F_{1/2}\left(\eta\right) + \beta F_{1/2}'\left(\eta\right)} = \frac{\beta}{F_{1/2}\left(\eta\right)} F_{1/2}'\left(\eta\right) \, . \tag{3.11}$$

If the Fermi energy ε_{F} is much larger than $k_{\mathrm{B}}T$, it follows that $k_{\mathrm{B}}T/\varepsilon_{\mathrm{F}} \to 0$. In this case the Fermi integrals are approximated by [see (2.38)]

$$F_{1/2}\left(\eta\right) \simeq \frac{2}{3}\eta^{\frac{3}{2}} \quad . \tag{3.12}$$

With this relation one easily obtains the derivatives with respect to η. The relative magnetization thus becomes

$$\zeta = \frac{\mu_{\mathrm{B}}H}{k_{\mathrm{B}}T}\frac{3}{2}\frac{k_{\mathrm{B}}T}{n\mu_{\mathrm{B}}} = \frac{M}{n\mu_{\mathrm{B}}} \quad ,$$

$$\Rightarrow \chi = \frac{M}{H} = \frac{3}{2}\frac{n\mu_{\mathrm{B}}^2}{\varepsilon_{\mathrm{F}}} \quad . \tag{3.13}$$

In (3.13) a new quantity has been introduced, namely the susceptibility χ. This quantity describes the response of the system to an applied magnetic field. In the simple (linearized) case χ is simply given by the ratio of M to H, in general χ is calculated from the derivative $\mathrm{d}M/\mathrm{d}H$. By expressing the Fermi energy in terms of the density of states (for one spin direction) of the free electron gas using

$$\mathcal{N}\left(\varepsilon_{\mathrm{F}}\right) = \frac{3}{4}\frac{n}{\varepsilon_{\mathrm{F}}} \tag{3.14}$$

χ takes a well known form, namely the Pauli susceptibility χ_{P} of the free electron gas

$$\chi = 2\mu_B^2 \mathcal{N}(\varepsilon_F) = \chi_P \quad . \tag{3.15}$$

It should be noted that the fact that the electrons in a metal are not free but move in a periodical potential which causes a correction to the bare Pauli susceptibility which has been introduced by Landau [48], Peierls [49], and Wilson [50] leading to

$$\chi = 2\mu_B^2 \mathcal{N}(\varepsilon_F) \left(1 - \frac{1}{3}\left(\frac{m}{m^*}\right)^2\right) \quad . \tag{3.16}$$

It is thus found that the application of an external field leads to a diamagnetic contribution of $-\frac{1}{3}$ of the Pauli susceptibility. The additional term $\frac{m}{m^*}$ accounts for the influence of the band structure which is embodied in the effective mass of the conduction electron m^* (see Chap. 4). In cases where the effective mass becomes small, the diamagnetic part can outweigh the paramagnetic part leading to an overall diamagnetic behavior of the metal (e.g. in bismuth). However, comparing the calculated and the experimental susceptibilities leads to a disappointing result. For fcc palladium, for instance, the observed value of χ is about 10 times higher than the calculated one. Similar discrepancies are observed for the transition metals but also for the simple alkali metals (see Table 3.1) which are actually assumed to show a free electron like behavior.

Table 3.1. Ratio of the experimental susceptibility χ_{expt} and the free electron susceptibility χ_P from (3.15)

Metal	$\chi_{\text{expt}}/\chi_P$	Ref.
Li	2.5	[51]
Na	1.67	[52]
K	1.51	[53]
Rb	1.60	[54]
Cs	1.74	[54]

It becomes obvious that the description of the free electron gas used up to now is not sufficient; what is missed are the effects of the electron–electron interaction, in particular the effect of the exchange interaction, which will provide the necessary correction.

The temperature dependence of the susceptibility can be calculated in analogy to the specific heat

$$\chi = \chi_0 \left(1 + aT^2\right) \quad \text{with} \quad a = \frac{\pi^2}{6} k_B^2 \left(\frac{\mathcal{N}(\varepsilon_F)''}{\mathcal{N}(\varepsilon_F)} - \left(\frac{\mathcal{N}(\varepsilon_F)'}{\mathcal{N}(\varepsilon_F)}\right)^2\right) . \tag{3.17}$$

Again, the temperature dependence is related to the details of the density of states at the Fermi energy. A detailed derivation of the coefficient a can

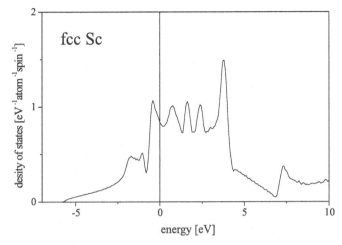

Fig. 3.1. Total density of states of fcc Sc. The Fermi energy is almost in a valley of the DOS, a position which can lead to an increasing susceptibility in a limited temperature range

be found in Sect. B. This relation can be used to define a temperature T_F given by $T_F^2 = a^{-1}$. This characteristic temperature (which depends on the peculiarities of the density of states at the Fermi energy) allows one to write the susceptibility in a concise form

$$\chi = \chi_0 \left(1 \pm \frac{T^2}{T_F^2} \right) \quad ,$$ (3.18)

which will be used again later on.

The expression for the Pauli susceptibility (3.15) should be compared with the result for the specific heat of the free electron gas (2.44). Both quantities describe the response of the electron gas to an excitation. For the specific heat it is the temperature and for the susceptibility it is the magnetic field. For the gas of free electrons, which consists of fermions, only those electrons which are on the surface of the Fermi sphere can be excited, which causes the proportionality to the density of states at the Fermi energy.

To study the influence of the peculiarities of the DOS at ε_F one has to distinguish between three fundamental cases which occur only in the low temperature regime:

1. $a > 0$, $\Rightarrow \varepsilon_F$ occurs at a minimum of the density of states, e.g. in Sc (Fig. 3.1); χ rises.
2. $a < 0$, $\Rightarrow \varepsilon_F$ occurs at a maximum of the density of states, e.g. in non-magnetic bcc Cr (Fig. 3.2); χ drops more strongly than $1/T$.
3. a more general behavior of χ as e.g. in Pd [104] or YCo_2 [55] requires higher terms of the Sommerfeld expansion. In the case of Pd and YCo_2 one finds a maximum in the susceptibility (Fig. 3.3).

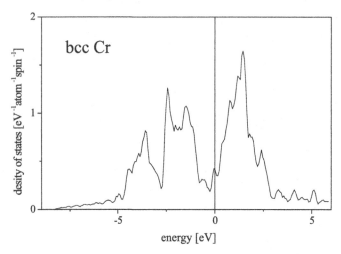

Fig. 3.2. Total density of states of bcc Cr. The Fermi lies in a local maximum of the DOS, a position which can lead to an stronger than usual decrease of the susceptibility in a limited temperature range

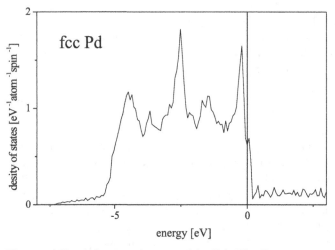

Fig. 3.3. Total density of states of fcc Pd. The Fermi energy is in a valley of the DOS, a position which indeed leads to an increasing susceptibility below about 90K

4. Energy Bands in the Crystal

In the preceding chapters it was assumed that the non-interacting electrons form a gaseous state moving freely in space as independent particles. This description is equivalent to the assumption of a constant potential in the respective Schrödinger equation so that the electron wavefunctions are represented by plane waves. For the free electrons the Schrödinger equation is thus of the form

$$\left[-\frac{\hbar^2}{2m} \nabla^2 + V(r) \right] \psi(r) = \epsilon \psi(r) \quad , \tag{4.1}$$

with $V(r) = 0 = $ constant. In a realistic solid the potential is far from being constant, but shows strong periodic oscillations caused by the Coulomb potential of the atomic cores and of the lattice periodicity of the electron density. As a contrast to the free electron description the model of tightly bound electrons (tight binding approximation) is introduced. This model is based on the assumption that a solid can be thought of as a periodic array of neutral atoms. The interaction of these atoms with one another should only be a small perturbation compared to the interaction between the various particles (electron and protons) inside the respective atoms. As a basis for a perturbational description the electron wavefunctions are taken from the free atom. The only necessary additional change to the wavefunctions stems from the translational symmetry of the ideal crystal by assuming a translation invariance of the electron density leading to

$$\psi = \psi_{\boldsymbol{k}}(r) = \sum_l e^{i\boldsymbol{k}R_l} \phi(r - R_l) \quad , \tag{4.2}$$

whereby $e^{i\boldsymbol{k}R_l}$ represents a plane wave (often called a phase factor) being the eigenfunction of the translation operator and $\phi(r - R_l)$ is the atomic wavefunction (Wannier-function) centered at R_l. Figure 4.1 clarifies the geometry assumed.

In quantum chemistry the expression for the wavefunction given in (4.2) is also known as an LCAO wavefunction (Linear Combination of Atomic Orbitals) since the total wavefunction is assumed to be a sum of atomic wavefunctions centered at the atomic positions R_l.

To describe the crystal potential the model sketched in Fig. 4.2 is applied. $U(r)$ is the unperturbed Coulomb potential of the single free atom. $V(r)$ is

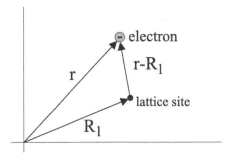

Fig. 4.1. Geometry for the tight binding model

the actual crystal potential, which is constructed by the superposition of the individual atomic potentials.

The Wannier-functions (atomic wavefunctions) are the solutions of the free atom Schrödinger equation

$$H_0 \phi(\mathbf{r}) = \left[-\frac{\hbar^2}{2m} \nabla^2 + U(\mathbf{r}) \right] \phi(\mathbf{r}) = E_0 \phi(\mathbf{r}) \quad . \tag{4.3}$$

Conversely the crystal Hamiltonian is of the form

$$H = \left[-\frac{\hbar^2}{2m} \nabla^2 + V(\mathbf{r}) \right] \quad . \tag{4.4}$$

Due to the lattice periodicity, one has to satisfy (4.3) also for the isolated atom with H replaced by

$$H_l = \left[-\frac{\hbar^2}{2m} \nabla^2 + U(\mathbf{r} - \mathbf{R}_l) \right] \quad , \tag{4.5}$$

for all l. To improve the legibility of the following equations the obvious spatial dependencies are omitted so that $V(\mathbf{r}) = V$, etc. By means of "resolution of the identity" one rewrites the crystal Hamiltonian

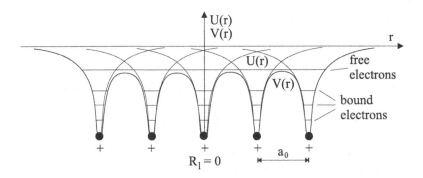

Fig. 4.2. Assumed potentials in the tight binding model, after [56]

$$H = H_l + (H - H_l) = H_l + (V - U) \quad . \tag{4.6}$$

Calculating the expectation value of the crystal Hamilton operator for the assumed wavefunction (4.2) one obtains

$$E = \frac{1}{N} \int \psi_{\boldsymbol{k}}^* H \psi_{\boldsymbol{k}} d\tau = \frac{1}{N} \left(\int \psi_{\boldsymbol{k}}^* H_l \psi_{\boldsymbol{k}} d\tau + \int \psi_{\boldsymbol{k}}^* (V - U) \psi_{\boldsymbol{k}} d\tau \right). \tag{4.7}$$

In (4.7) N is the norm of the wavefunction: $N = \int \psi_{\boldsymbol{k}}^* \psi_{\boldsymbol{k}} d\tau$.

For the eigenvalue of H_l with respect to $\psi_{\boldsymbol{k}}$ one finds, using (4.2)

$$H_l \psi_{\boldsymbol{k}} = \sum_l e^{i\boldsymbol{k}\boldsymbol{R}_l} H_l \phi (\boldsymbol{r} - \boldsymbol{R}_l) = E_0 \psi_{\boldsymbol{k}} \quad , \tag{4.8}$$

and hence

$$\begin{aligned}
E &= E_0 + \frac{1}{N} \int \psi_{\boldsymbol{k}}^* (H - H_l) \psi_{\boldsymbol{k}} d\tau \\
&= E_0 + \frac{1}{N} \sum_m \sum_l e^{i\boldsymbol{k}(\boldsymbol{R}_l - \boldsymbol{R}_m)} \int \phi^* (\boldsymbol{r} - \boldsymbol{R}_l) (V - U) \phi (\boldsymbol{r} - \boldsymbol{R}_m) d\tau \\
&= E_0 + \frac{1}{N} \sum_{\boldsymbol{R}=0, nn} e^{i\boldsymbol{k}\boldsymbol{R}} \int \phi^* (\boldsymbol{r} - \boldsymbol{R}) (V - U) \phi (\boldsymbol{r}) d\tau \quad . \tag{4.9}
\end{aligned}$$

The energy eigenvalue for the crystal Hamiltonian is thus expressed by a large component E_0 which stems from the solution of the isolated atom and a small contribution (second term of (4.9)) which depends only on the difference between the free atom potential and the crystal potential. Via a transformation of the coordinates of the atomic positions $\boldsymbol{R} = \boldsymbol{R}_l - \boldsymbol{R}_m$ in (4.9) the general summation over all lattice sites is transformed into a summation starting from the atom at the site with $\boldsymbol{R} = 0$ and is carried out further over its nearest-neighbors (nn). In general this summation must however be taken over the whole crystal. However, by assuming atomic wavefunctions whose amplitude rapidly decreases to zero for increasing values of $\boldsymbol{r}$, this summation can often being reduced to the next nearest-neighbor shells only. The summation in (4.9) is now split up into two parts, one for $\boldsymbol{R} = 0$ (on site) and one over the next nearest-neighbors (off site). These involve two integrals which are abbreviated as A and B describing the crystal field effects and the electron hopping (overlap), respectively

$$-A = \frac{1}{N} \int \phi^* (\boldsymbol{r}) (V - U) \phi (\boldsymbol{r}) d\tau \quad , \tag{4.10}$$

$$-B = \frac{1}{N} \int \phi^* (\boldsymbol{r} - \boldsymbol{R}) (V - U) \phi (\boldsymbol{r}) d\tau \quad . \tag{4.11}$$

Assuming that ϕ is spherically symmetrical (e.g. an s-orbital), then; since $|\boldsymbol{r} - \boldsymbol{R}|$ is the same for all next nearest-neighbors, the hopping integral always has the same value, so that the only remaining summation is that over the phase factors. If one restricts the summation to a single nearest neighbor shell only the energy E becomes

$$E(\mathbf{k}) = E_0 - A - B \sum_{nn} e^{i\mathbf{k}\mathbf{R}} \quad . \tag{4.12}$$

For a simple cubic (sc) lattice with lattice constant a_0 one obtains for the s-electrons ($l = 0, m = 0$)

$$E(\mathbf{k}) = E_0 - A - 2B(\cos(a_0 k_x) + \cos(a_0 k_y) + \cos(a_0 k_z)) \quad . \tag{4.13}$$

Between the center of the Brillouin-zone (BZ) $\mathbf{k} = 0$ (Γ–point) and the surface of the first BZ the energy E varies between $E_0 - A \mp 6B$. A contour plot of the energy given by (4.13) in the k_x, k_y plane is shown in Fig. 4.3.

Fig. 4.4 depicts the s-band along the k_x direction between the center of the Brillouin zone ($\mathbf{k} = 0$) and the reciprocal lattice vector ($k_x = \frac{\pi}{a_0}$),(length of the Brillouin zone).

B depends on the distance $|\mathbf{r} - \mathbf{R}|$ so that external pressure will affect its value. As the overlap increases with increasing pressure also the bandwidth will increase accordingly. That means that the same number of electronic states is now dispersed over a wider energy range, making the density of states at the Fermi energy smaller. From this simple picture it is obvious that all quantities which rely on the density of states at the Fermi energy, like the specific heat or the susceptibility, will also be influenced by pressure.

In general, if the band dispersion $E(\mathbf{k})$ is given, the density of states (defined as the number of states in the energy unit) can be calculated according to

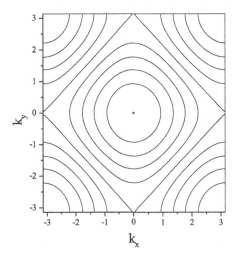

Fig. 4.3. Energy contours for the s-band in the tight binding model ($k_z = 0$). The lattice constant is assumed to be unity and the band energies are in arbitrary units

$$\mathcal{N}(\varepsilon) = \frac{1}{8\pi^2} \int (\nabla_{\boldsymbol{k}} E(\boldsymbol{k}))^{-1} \, dS_k \quad , \tag{4.14}$$

dS_k being an element of the constant energy surface $E(\boldsymbol{k}) = \varepsilon = $ const.

In the case of a small wave vector $\boldsymbol{k}$, (long wave length) (4.13) can be expanded in a Taylor series for the $\cos x$ which gives the well known parabolic relation for the free electron gas

$$E(\boldsymbol{k}) = E_0 - A - 6B + Ba^2 \boldsymbol{k}^2 \quad . \tag{4.15}$$

Comparing (4.15) with the solution for the free electron gas one can reach a formal identity by introducing the effective mass $m^* = \hbar^2/(2a^2 B)$. The meaning of the effective mass is that electrons moving in a band with low dispersion (small B) behave like heavier particles than those who are moving freely. This behavior has strong consequences on properties like the conductivity, the specific heat, the Hall effect, etc. As long as the electron momentum is small the behavior of the electrons in the crystal can be described by the free electron gas. Small momentum means long wavelength of the respective plane wave, and thus the electron feels only the mean value of the potential averaged over a large number of atomic sites.

For a realistic description of transition metals it is of course not sufficient to treat only s-electrons since in these metals the d-electrons – and the interaction between the s- and the d-electrons – are important. For d-electrons one has to consider the spatial and angular dependence of the five orbitals. According to the magnetic quantum number m the atomic wavefunctions are fivefold degenerate. In a crystal (with cubic or tetragonal symmetry) one usually uses symmetry adapted wavefunctions which are linear combinations of the atomic ones (see also Sect. C.). Using the Cartesian representation these five orbitals read

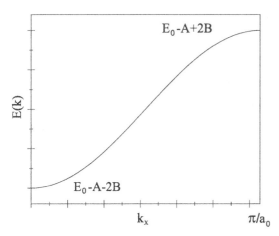

Fig. 4.4. One dimensional dispersion along k_x for the s-band in the tight binding approximation

$$\phi_1 = C_1 xy \frac{f(r)}{r^2} \quad \phi_2 = C_2 xz \frac{f(r)}{r^2} \quad \phi_3 = C_3 yz \frac{f(r)}{r^2}$$

$$\phi_4 = C_4 \left(x^2 - y^2\right) \frac{f(r)}{r^2} \quad \phi_5 = C_5 \left(3z^2 - r^2\right) \frac{f(r)}{r^2} \quad . \tag{4.16}$$

The functions ϕ_1, ϕ_2, ϕ_3 are very often named t_{2g}-orbitals, the remaining ϕ_4, ϕ_5 are known as e_g-orbitals. The crystal wavefunction is thus given by a linear combination of the orbital wavefunctions

$$\psi_{n\boldsymbol{k}}(\boldsymbol{r}) = \sum_{m=1}^{5} a_{nm}(\boldsymbol{k}) \Phi_{m\boldsymbol{k}}(\boldsymbol{r}) \quad , \tag{4.17}$$

with

$$\Phi_{m\boldsymbol{k}}(\boldsymbol{r}) = \sum_{l} e^{i\boldsymbol{k}\boldsymbol{R}_l} \phi_m (\boldsymbol{r} - \boldsymbol{R}_l) \quad . \tag{4.18}$$

Determining the coefficients $a_{nm}(\boldsymbol{k})$ by applying the variational principle leads to a matrix eigenvalue problem of the type

$$|H_{nm} - E\delta_{nm}| = 0 \quad , \tag{4.19}$$

$$H_{nm} = E_0 \delta_{nm} + \frac{1}{N} \sum_{\boldsymbol{R}=0,nn} e^{i\boldsymbol{k}\boldsymbol{R}} \int \phi_n^* (\boldsymbol{r} - \boldsymbol{R}) (V - U) \phi_m(\boldsymbol{r}) \, d\tau , \tag{4.20}$$

and, since the ϕ_n on the same atom are orthonormal, the on-site terms can be found easily and are put into a quantity C

$$H_{nm} = (E_0 + C) \delta_{nm} + \frac{1}{N} \sum_{nn} e^{i\boldsymbol{k}\boldsymbol{R}} \int \phi_n^* (\boldsymbol{r} - \boldsymbol{R}) (V - U) \phi_m(\boldsymbol{r}) \, d\tau \tag{4.21}$$

The matrix elements H_{nm} depend on the hopping integrals and on trigonometric functions. For cubic symmetry it is possible to construct three independent integrals the so called Fletcher integrals.

As mentioned above, the electronic properties of transition metals can only sufficiently be described by considering at least s- and d-wavefunctions. The matrix eigenvalue problem then becomes of block diagonal form. The eigenvalues are determined from the nodes of the secular equation which is given by the determinant being set to zero:

$$\begin{vmatrix} s & sd - \text{hybrid} \\ sd - \text{hybrid} & (H_{nm} - E\delta_{nm}) \end{vmatrix} = 0 \quad . \tag{4.22}$$

All modern methods (and the respective computer programs) to determine the electronic band structure of solids essentially follow these ideas. They differ in the choice of the basis set which is very often tailor-made for the problem to be solved. An excellent introduction into the subject of electronic band structure calculations and their application to magnetic systems can be found in [146].

5. Experimental Basis of Ferromagnetism

This chapter provides an overview of experimental findings concerning the magnetic properties of ferromagnetic systems, in particular of Fe, Co, Ni, and their alloys. These experimental results provide the basis for all theoretical models, whether phenomenological or from first principles (ab initio). However, before going into the details of the experiment one has to define what to measure.

In vacuo the magnetic field $\boldsymbol{H}$ and the induction $\boldsymbol{B}$ are related via:

$$\boldsymbol{B} = \mu_0 \boldsymbol{H} \quad \text{(SI)} \tag{5.1}$$

where μ_0 is the vacuum permeability and has the value $\mu_0 = 4\pi \times 10^{-7} \frac{\text{Vs}}{\text{Am}}$. To describe the magnetic state of matter one introduces the magnetization $\boldsymbol{M}$ so that the total induction becomes

$$\boldsymbol{B} = \mu_0 \left(\boldsymbol{H} + \boldsymbol{M} \right) \quad . \tag{5.2}$$

The magnetization is equal to the density of the magnetic dipoles $\boldsymbol{m}$

$$\boldsymbol{M} = \boldsymbol{m} \frac{N}{V} \quad . \tag{5.3}$$

To make things a bit less confusing one introduces an external induction $\boldsymbol{B}_0$ to replace the external field so that (5.2) reads

$$\boldsymbol{B} = \boldsymbol{B}_0 + \mu_0 \boldsymbol{M} \quad . \tag{5.4}$$

It is found that very often there exists a linear relation between the external induction and the magnetization in the specimen. The proportionality constant is the susceptibility χ and one writes

$$\mu_0 \boldsymbol{M} = \chi \boldsymbol{B} \quad . \tag{5.5}$$

If χ is negative the induced magnetic polarization opposes the applied field. In this case one speaks of a *diamagnetic* behavior. In the case $\chi > 0$ a paramagnetic behavior is found so that the induced magnetic polarization acts in the same direction as the applied field and thus enhances it. In general the susceptibility of atoms is composed of a diamagnetic χ_{dia} and a paramagnetic χ_{para} contribution to χ. The paramagnetic part is due to the orientation of already present intrinsic magnetic moments by the applied field. These magnetic moments stem from the orbital angular momenta of the electrons and from their spin. The magnetic dipole moment due to the orbital angular momentum of an electron is given by

$$\boldsymbol{m} = -\frac{e}{2m_{\mathrm{e}}} \sum_i \boldsymbol{r}_i \times \boldsymbol{p}_i = -\mu_{\mathrm{B}} \boldsymbol{L} \,, \tag{5.6}$$

where $\mu_{\mathrm{B}} = \frac{e\hbar}{2m_{\mathrm{e}}}$ is the *Bohr magneton* and $\boldsymbol{L}$ is the (operator of the) orbital angular momentum defined by

$$\boldsymbol{L} = \frac{1}{\hbar} \sum_i \boldsymbol{r}_i \times \boldsymbol{p}_i \,. \tag{5.7}$$

In addition to the angular momentum the electrons possess a spin which can be treated as an angular momentum. The sum of the individual electron spins yields the total spin-momentum of the atom and reads

$$\boldsymbol{m} = \mu_{\mathrm{B}} g_{\mathrm{s}} \sum_i \boldsymbol{s}_i = \mu_{\mathrm{B}} g_{\mathrm{s}} \boldsymbol{S} \,, \tag{5.8}$$

where g_{s} is the electron g-factor ($g_{\mathrm{s}} = -2.0023$). Both $\boldsymbol{L}$ and $\boldsymbol{S}$ can also be seen as quantum mechanical operators which allows a direct evaluation for quantum mechanical systems.

For closed shells both the orbital angular momentum and the spin momentum adds up to zero. Only open shells contribute to magnetic phenomena. Typical examples are the open d-shells in the 3d-atoms or the 4f-shell of the rare earths. For both cases we can expect a paramagnetic behavior.

In addition to the paramagnetic contribution there is also the diamagnetism of the electrons. It is reasoned by the classical electrodynamical effect that a magnetic field causes a circular current which itself produces a magnetic field which is opposed to the inducing one. In classical electrodynamics this effect is called *Lenz's rule*. Due to this effect the susceptibility always contains a negative diamagnetic contribution. The usual treatment of the diamagnetism of atoms and ions employs the *Larmor theorem*.

- In a magnetic field the motion of electrons around a central nucleus is, to the first order in $\boldsymbol{B}$, the same as a possible motion in the absence of $\boldsymbol{B}$ except for the superposition of a precession of the electrons with the frequency

$$\omega = \frac{e\boldsymbol{B}}{2m_{\mathrm{e}}} \quad (\mathrm{SI}) \,. \tag{5.9}$$

If the average electron current around the nucleus is zero initially, the application of the magnetic field will cause a finite current around the nucleus. The current is equivalent to a magnetic moment opposite to the applied field. It is assumed that the Larmor frequency (5.9) is much lower than the original motion in the central field. The Larmor precession of Z electrons is equivalent to an electric current I (charge×revolutions per unit time)

$$I = -Ze \left(\frac{1}{2\pi} \frac{e\boldsymbol{B}}{2m_{\mathrm{e}}} \right) \quad . \tag{5.10}$$

The magnetic moment $\boldsymbol{m}$ of a current loop is given by the product (current$\times$ area of the loop). The area of a loop of radius ρ is simply $\rho^2 \pi$ giving

$$\boldsymbol{m} = -\frac{Ze^2 \boldsymbol{B}}{4m_{\mathrm{e}}} \langle \rho^2 \rangle \quad \text{(SI)}. \tag{5.11}$$

where $\langle \rho^2 \rangle = \langle x^2 \rangle + \langle y^2 \rangle$ is the mean square of the perpendicular distance of the electron from the field axis through the nucleus. The mean square distance of the electrons from the nucleus is $\langle r^2 \rangle = \langle x^2 \rangle + \langle y^2 \rangle + \langle z^2 \rangle$. For a spherically symmetrical distribution of charge we have $\langle x^2 \rangle = \langle y^2 \rangle = \langle z^2 \rangle$ so that $\langle r^2 \rangle = \frac{2}{3} \langle \rho^2 \rangle$. With $n = \frac{N}{V}$ one obtains

$$\chi_{\mathrm{dia}} = -\frac{\mu_0 n Z e^2}{6m_{\mathrm{e}}} \langle r^2 \rangle \tag{5.12}$$

which is the classical result for the *Langevin* susceptibility. Typical examples for diamagnetic systems are the inert gases but also metals with closed electron shells.

Experimentally the magnetic moment is usually given in units of emu/g, emu/cm^3 or emu/mole. In theoretical papers it is more common to measure the moment as the number of unpaired spins, leading to a magnetic moment M or in units of μ_{B} (Bohr magnetons). A collection of various magnetic units and their conversion between the different units systems can be found in Sect. K.

$$1\mu_{\mathrm{B}} = \frac{e\hbar}{2m_{\mathrm{e}}}$$
$$= 5.78838263 \times 10^{-5} \mathrm{eV/T}$$
$$= 9.27401543 \times 10^{-24} \mathrm{J/T} \quad .$$

Table 5.1 gives the experimental magnetic moment σ and the Curie temperature T_{c} for the three ferromagnetic transition metals

Table 5.1. Magnetic moments of the ferromagnetic $3d$ transition metals. Note that cobalt which has hcp structure at $T = 0\mathrm{K}$ is fcc at T_{c}

	σ [emu/g]	σ [μ_{B}]	T_{c} [K]	ρ at 298K $\left[\mathrm{g/cm}^3\right]$
Fe (bcc)	221.7	2.22	1044	7.875
Co (fcc)	166.1	1.75	1388	8.793
Co (hcp)	163.1	1.72	1360	8.804
Ni (fcc)	58.6	0.62	627	8.912

Plotting the magnetic moment σ of the pure metals and of the binary alloys as a function of the electron concentration n_e (number of valence electrons per atom) for the $3d$ transition metals yields the Slater–Pauling (S-P) curve [57] (Fig. 5.1).

According to the position on the S-P curve one makes a phenomenological distinction into two classes of systems which are given by their position on the left (ascending) or right (descending) branch:

- weakly ferromagnetic.... $\frac{d\sigma}{dn} > 0$ ascending branch.
- strongly ferromagnetic... $\frac{d\sigma}{dn} < 0$ descending branch.

The magnetic moments of the transition metals are basically different from the moments of free atoms, ions or atoms bound in chemical complexes (or of the rare earth metals) as they are never given by an integer number of unpaired spins. This behavior is typical for itinerant systems and is a well known deviation from Hund's rules. For a review of Hund's rules see Sect. I. In contrast to the itinerant electron systems, those with localized magnetic moments can readily be understood from their normal quantum state. Their state is defined by the quantum numbers L, S, and J. The total magnetic moment (in atomic physics traditionally for these moments the letter μ is used) of such an atom consists of an orbital and a spin contribution given by

$$\mu_l = g_l \mu_B \frac{L}{\hbar} \quad , \quad \mu_s = g_s \mu_B \frac{S}{\hbar} \quad , \tag{5.13}$$

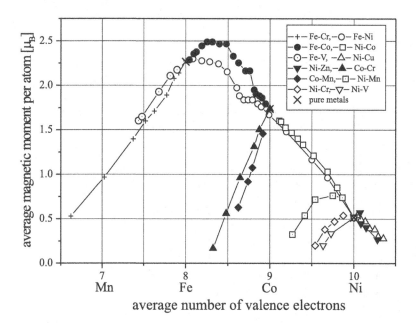

Fig. 5.1. Slater–Pauling curve (after [57])

where $g_l = -1$ and $g_s = -2.0023$ are the respective gyromagnetic factors. Due to the spin-magnetic anomaly (from here on g_s will be approximated by ($g_s = -2$) whenever appearing throughout the whole book) the total magnetic moment is not simply the sum of the angular and the spin component (given by J) but is given by multiplication by the Landé factor g_j

$$\mu_j = g_j \mu_B \frac{J}{\hbar} \quad , \qquad \text{with} \tag{5.14}$$

$$g_j = -\frac{3J(J+1) - L(L+1) + S(S+1)}{2J(J+1)} \quad . \tag{5.15}$$

This description usually holds well for free atoms, ions in chemical complexes or localized electron states like the $4f$ electrons in the rare earth elements. In the rare earth metals the magnetic moment is provided by $4f$ electrons, which are shielded by the $6d$ electrons and which are not involved in the chemical bonding and are thus localized (atomic like), the magnetic moments given by Hund's rules thus agree very well with experiment.

In solids the angular moment appears to be always close to zero, an effect which is called the quenching of the orbital momentum. The reason for this effect is that in solids the crystal-field of the neighboring atoms is much larger that the spin orbit splitting (which gave rise to the ordering due to Hund's rules) and thus J and L are no longer good quantum numbers (see Sect. C.). The magnetic moment is thus given only by the value of the spin-moment. For the light elements ($Z < 49$, indium) the orbital moment is usual negligible compared with the spin contribution, as the angular momentum appears to be quenched in these systems. For the heavier elements, when spin-orbit coupling becomes more important (which is due to relativistic effects becoming progressively more important for the heavier elements) the orbital moment recovers and must be taken into account. However also for these systems very often a scalar relativistic treatment is sufficient and the $\boldsymbol{L} \cdot \boldsymbol{S}$-coupling interaction can be accounted for in a perturbational treatment. The respective correction to the Hamiltonian is of the form

$$W_{LS} = \frac{1}{2m^2 c^2} \frac{1}{r} \frac{dV}{dr} \boldsymbol{L} \cdot \boldsymbol{S} = \Gamma(r) \boldsymbol{L} \cdot \boldsymbol{S} \quad , \tag{5.16}$$

$\Gamma(r)$ is the Thomas-factor which becomes small for the weakly bound valence electrons where r is large and because of the flat potential dV/dr is also small.

The breakdown of Hund's rules in a crystalline solid is easily seen by applying them to the magnetic $3d$ transition metals. For Fe, Co and Ni the magnetic spin-moment due to Hund's rules would be 4, 3, and $2\mu_B$ respectively. These numbers neither resemble the experimental trend for the moments per atom in the solids nor do they come close to the absolute numerical values (see Table 5.1).

5.1 Nickel Alloys

In most Ni-rich alloys of the form $Ni_{1-c}X_c$, σ and T_c vary in an astonishingly simple manner as a function of the concentration c:

$$\sigma(c) = \sigma(0)(1 - \nu c) \quad , \quad T(c) = T(0)(1 - \nu c) \quad , \tag{5.17}$$

where ν appears as a kind of magnetic valence: $\nu = 1, 2, 3, 4, 6$ for Cu, Zn, Al, Ti, Cr. (Fig. 5.2) A more rigorous formulation of the concept of magnetic valence is given in Chap. 9

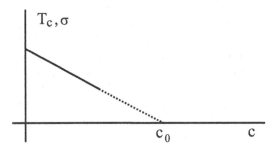

Fig. 5.2. Concentration dependence of the Curie temperature T_c and the magnetization σ in Ni alloys

However, when the susceptibility becomes very large, or the magnetic moments become very small, there are deviations from this simple law. Examples are Ni alloys with palladium and platinum ($Ni_{1-c}Pt(Pd)_c$: $c_0 \simeq 58\%$), (Fig. 5.3) and Ni-Al alloys ($Ni_{1-c}Al_c$: $c_0 \simeq 25\%$) in particular the very weak ferromagnet Ni_3Al (Fig. 5.4).

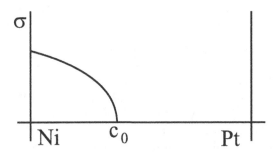

Fig. 5.3. Concentration dependence of the magnetization σ in Ni-Pt alloys

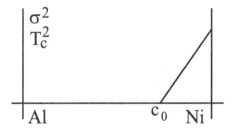

Fig. 5.4. Concentration dependence of the Curie temperature T_c and the magnetization σ in Ni-Al alloys. The way of plotting T_c^2 or σ^2 as a function of the concentration is sometimes referred to as *Mathon plot*

5.2 Iron Alloys

These alloys more or less show a rather uniform behavior. In most cases the average magnetic moment scales with the concentration and is given by the relation: $\sigma(c)/c \simeq 2.2\mu_B$. This increasing moment behavior is typical for weak ferromagnets which, as a consequence of alloying are gradually transformed into a strong ferromagnet.

As an example for an iron alloy with a magnetic partner the system Fe-Co is shown (Fig. 5.5). While the moment at the Co site stays constant as a function of the concentration, the iron moment increases, following approximately the S-P curve.

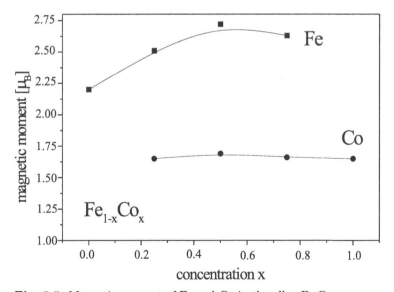

Fig. 5.5. Magnetic moment of Fe and Co in the alloy Fe-Co

5.3 Palladium Alloys

Palladium is a most interesting system. Not only does its susceptibility deviate strongly from the Pauli value of the free electron gas, but it is almost ferromagnetic so that very small amounts of $3d$ transition metal impurities cause a ferromagnetic transition [58]. As the susceptibility describes the magnetic polarizability of a system, one concludes that Pd is on the verge of a magnetic instability so that these magnetic impurities drive it into magnetic order. Historically, however, it was assumed that the magnetic moment measured is located only at the $3d$ transition metal atoms. This interpretation yielded magnetic moments at the impurities up to $10\mu_B$ per impurity atom, an effect described as "giant moments". It is clear that such systems have raised the interest of theoreticians as well as of experimentalists. Both from calculations and from neutron diffraction it was seen that the impurity moments are actually limited to their number of holes in the d-band which gives about $4, 3, 2$ μ_B for Mn, Fe, Co respectively. The giant moment stems from the large polarization cloud around these impurity atoms. Depending on the impurity concentration, the number of magnetically polarized Pd-atoms inside this cloud can be as large as 1000 atoms [160]. Adding up all these small Pd moments and attributing this total moment to the impurity transition metal explains the giant moment.

5.4 Iron–Nickel Alloys

The alloy system Fe-Ni exhibits extraordinary properties. As iron crystallizes in the bcc lattice and Ni in the fcc lattice, the alloy system shows a phase transition from the α (bcc) to the γ (fcc) phase for a Ni concentration of about 32%. Close to this transition is the alloy $Fe_{65}Ni_{35}$ which shows the so called Invar behavior. The name Invar means that this alloy does not show any thermal expansion around ambient temperature. The Invar effect was discovered in 1897 by the Swiss-born scientist C.E. Guillaume [59], who was presented with the 1920 physics Nobel prize for his work on the metallurgy of Invar systems. The reason for the vanishing thermal expansion (see Fig. 5.6) is a compensation between the *positive* phonon part and the *negative* magneto-volume part of the thermal expansion coefficient. Apart from the vanishing thermal expansion there exist a number of other effects which are related to the Invar anomaly like a maximum in the susceptibility, a maximum in the atomic volume and a minimum in the bulk modulus.

5.5 Effects of Strong Magnetic Fields

If one applies an external magnetic field to a system, the magnetic moment of the system varies according to the susceptibility. In general the relation

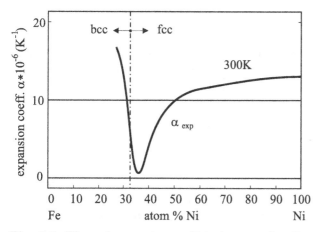

Fig. 5.6. Thermal expansion coefficient α as a function of the concentration in Fe-Ni. The dashed-dotted line denotes the phase boundary between the α- (bcc) and the γ-phase (fcc)

is linear. Deviations from this linearity can either be caused by higher order terms related to the itinerant electrons or by spin fluctuations or excitation processes (Schottky anomalies); other reasons might include metallurgical effects such as gradients in the alloy concentrations.

An elegant way to plot the results of high field experiments was introduced by A. Arrott and is commonly known as an Arrott plot (Fig. 5.7), where the axes are scaled in terms of M^2 and H/M

If the resulting graphs are straight lines, one obviously can write M^2 in the following form

$$M^2 = M_0^2 + b\frac{H}{M} \quad ,$$

$$\Rightarrow F = A\frac{M^2}{2} + B\frac{M^4}{4} \quad , \tag{5.18}$$

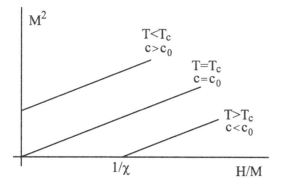

Fig. 5.7. Arrott plot

$$A = -\frac{M_0^2}{b} \quad , \quad B = \frac{1}{b} \quad .$$

This phenomenological form of the free energy F is not immediately obvious. It will become clear when the magnetic phenomena are described within the Landau theory of phase transitions (Chap. 14).

5.6 Effects of High Pressure

Pressure experiments provide a lot of information about the magnetic properties. For any itinerant ferromagnet there exists a critical pressure for the disappearance of ferromagnetism, at which these systems are transformed into a real non-magnetic state. This non-magnetic state has to be distinguished from the paramagnetic state above T_c, being truly non-magnetic so that no microscopic magnetic moments exists. From thermodynamics one obtains the relation between the pressure dependence of the magnetic moment σ, the Curie temperature T_c and the susceptibility χ

$$\frac{d \ln \sigma}{dP} = \lambda \frac{d \ln T_c}{dP} = -\frac{\lambda}{2} \frac{d \ln \chi}{dP} \quad . \tag{5.19}$$

The coefficient λ depends on the underlying model and its value allows a classification of the magnetic interaction present in the system studied.

In an Arrott plot pressure leads to a parallel shift of respective isotherms (Fig. 5.8 a) .

For Fe-Ni, Fe-Pd, and Fe-Pt one finds a very simple relation for the change of the Curie temperature with pressure

$$\frac{dT_c}{dP} = -\frac{\alpha}{T_c} \quad ,$$

$$\Rightarrow T_c^2 (P) = T_c^2 (0) \left(1 - \frac{P}{P_c} \right) \quad , \tag{5.20}$$

$$P_c = \frac{T_c^2 (0)}{2\alpha} \quad ,$$

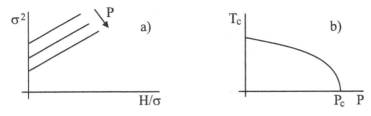

Fig. 5.8. (a) Arrott plot for the external pressure as parameter, (b) pressure dependence of the Curie temperature T_c according to (5.20)

where P_c is the critical pressure for the disappearance of magnetism (Fig.5.8 b). It should be noted again that above the critical pressure, a system becomes truly non-magnetic. This state is different from the paramagnetic state which appears above the critical temperature T_c which is due to a breakdown of long range order between the spins. This is the reason why pressure experiments are most useful to distinguish between these two macroscopically "non-magnetic" but microscopically essentially different states.

There are deviations from the simple behavior described by (5.20). For pure Fe and Co dT_c/dP is zero (or at least very small) over a wide pressure range and for pure Ni dT_c/dP is even slightly positive for small pressure. However, this does not mean that these metals no critical pressure P_c exists. The slightly positive pressure dependence found for Ni in the low pressure range is due to its peculiar electronic structure.

5.7 Effects of Finite Temperature

Up to this point it has simply been assumed that there exists a critical temperature where magnetism disappears, which is called the Curie temperature. Experimentally it is found that the bulk magnetic moment of a sample goes to zero which in the framework of the Arrott plot means that the M^2 graph passes through the origin, so that for $H/M = 0$ also $M^2 = 0$. This macroscopic picture does not allow us to say anything about what is going on at the individual atom and its individual moment. This question about a local moment will be discussed later. The Curie temperature is a macroscopic quantity which is given by a pole in the susceptibility and concomitantly the magnetic moment goes to zero. Depending on the system and on the theoretical model assumed, various analytical relations temperature variation of the magnetic moment $\sigma(T)$ are found which can be written in the general form

$$\frac{\sigma(T)}{\sigma(0)} = X \quad . \tag{5.21}$$

In (5.21) depending on the underlying model, X stands for:

- first and second order spinwaves: $X = 1 - aT^{3/2} + bT^{5/2}$
- Stoner excitations (Fermi liquid): $X = 1 - cT^2$
- energy gap Δ at ε_F: $X = 1 - dT^n \exp\left(-\frac{\Delta}{k_B T}\right)$
- Weiss model: $X = B(a, J)$.
- Spin fluctuations: $X = \sqrt{1 - eT}$

5.8 Susceptibility above T_c

As the susceptibility $\chi(T)$ diverges at T_c, it is often more convenient to argue in terms of the node of the reciprocal susceptibility $1/\chi(T_c)$. One finds that for most systems (and at sufficiently high temperature) $\chi(T)^{-1}$ varies approximately linearly with T

$$\frac{1}{\chi(T)} = C(T - T_{\mathrm{c}}) \quad \text{for } T > T_{\mathrm{c}} \quad , \tag{5.22}$$

$$C = \frac{\mathrm{d}\left(\chi^{-1}\right)}{\mathrm{d}T} \quad \text{Curie constant.} \tag{5.23}$$

This behavior is called the Curie–Weiss law. Deviations from it are related to peculiarities of the systems or the models. This linear relation is a crucial test for the quality of a model. However, in the high temperature case, where one approaches the classical regime, the Curie–Weiss law is established in any case. At low temperatures however and around the Curie temperature deviations from the Curie–Weiss behavior are very common. Fig. 5.9 sketches a the large variety of cases found in real systems.

5.8.1 Susceptibility of "Classical Spins"

To derive the classical result one assumes an ensemble of N particles carrying a classical spin (moment) which in the absence of an external field are oriented at random. Each of these spins $\pm \frac{1}{2}$ thus carries a magnetic moment $\mu = \mp \frac{g_s \mu_{\mathrm{B}}}{2}$, (g_s is the gyromagnetic ratio for the spin and takes the value -2). If one applies an external field H one can calculate the resulting total magnetic moment by counting the number of particles who orient their moments parallel and antiparallel to the direction of the field where the energy associated with this orientation is given by $\pm \mu H$. The average magnetic moment per particle is then given by

$$m = \frac{M}{N} = \mu \frac{\exp\left(\frac{\mu H}{k_{\mathrm{B}} T}\right) - \exp\left(-\frac{\mu H}{k_{\mathrm{B}} T}\right)}{\exp\left(\frac{\mu H}{k_{\mathrm{B}} T}\right) + \exp\left(-\frac{\mu H}{k_{\mathrm{B}} T}\right)}$$

$$= \mu \tanh\left(\frac{\mu H}{k_{\mathrm{B}} T}\right) \quad . \tag{5.24}$$

The functional dependence involving the tanh given by (5.24) is identical with the Brillouin function for a two level system (see Chap. 6). Since the susceptibility is defined as the derivative of the magnetic moment with respect to the applied field one obtains

$$\chi = \frac{\mathrm{d}m}{\mathrm{d}H}$$

$$= \frac{\mu^2}{k_{\mathrm{B}} T} \frac{1}{\cosh^2\left(\frac{\mu H}{k_{\mathrm{B}} T}\right)} \quad . \tag{5.25}$$

Figure 5.10 shows this functional dependence which is characterized by an exponential increase for low temperatures and a $1/T$ behavior in the high

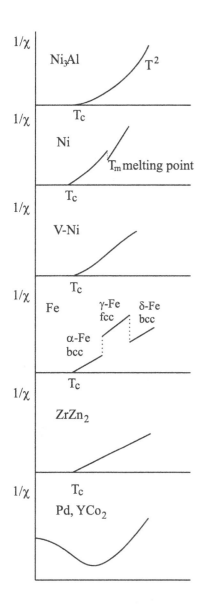

Fig. 5.9. A sketch of the different behavior found for the susceptibility above T_c in various systems

temperature range (Schottky behavior, Schottky anomaly). It should be noted that for the specific heat of a 2-level system exactly the same type of anomaly occurs and that the name "Schottky anomaly" was originally coined for this case.

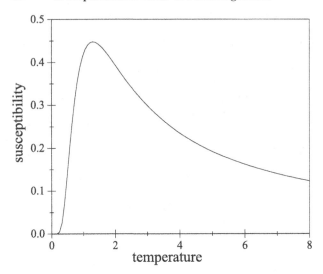

Fig. 5.10. Susceptibility of a two-level system. The increase in the low temperature region is about exponential, for high temperature the susceptibility shows a $1/T$ behavior. Susceptibility and temperature are given in arbitrary units

The high temperature behavior can be determined by rewriting (5.25) for the inverse susceptibility and expanding the hyperbolic cosine for small arguments (large T) so that

$$\chi^{-1}(T) = \frac{k_B T}{\mu^2} \left[1 + \frac{1}{2!} \left(\frac{\mu H}{k_B T} \right)^2 + \frac{1}{4!} \left(\frac{\mu H}{k_B T} \right)^4 + ... \right]. \tag{5.26}$$

For sufficiently large temperature the inverse susceptibility becomes linear in T and scales with a Curie constant of the value $C = \frac{4}{g_s{}^2 \mu_B^2}$. A more detailed evaluation of the properties of a "classical spin" system is given in Sect. D.

5.9 Critical Exponents

At the critical temperature the behavior of a thermodynamic function $F(t)$ is determined by its critical exponent, where t is the deviation from the critical temperature defined as [83]

$$t = \frac{T - T_c}{T_c} . \tag{5.27}$$

The critical exponent is then defined as

$$\lambda = \lim_{t \to 0} \frac{\ln |F(t)|}{\ln |t|} . \tag{5.28}$$

Since any function $F(t)$ can be written in the form

$$F(t) = A |t|^{\lambda} \left(1 + bt^{\lambda_1} + \ldots\right) \quad . \tag{5.29}$$

λ can always be calculated using (5.28). Experimentally one performs highly accurate measurements of the thermodynamic property of interest (specific heat, magnetization, susceptibility, etc.) and plots the experimental dependence around the critical point on a double logarithmic scale. From (5.28) it is obvious that λ is then given by the slope of the resulting graph at $t = 0$. In the literature the various critical exponents are characterized by specific Greek letters as there are:

$$\text{Zero-field specific heat} \quad c_H \sim |t|^{-\alpha} \quad ,$$

$$\text{Zero-field magnetization} \quad M \sim (-t)^{-\beta} \quad ,$$

$$\text{Zero-field isothermal susceptibility} \quad \chi_T \sim |t|^{-\gamma} \quad ,$$

$$\text{Correlation length} \quad \xi \sim |t|^{-\nu} \quad .$$

It is known that the critical exponents of a system are not independent of each other but obey certain scaling relations. The reason for this general behavior is that the exponents are to a large degree *universal* depending only on a few fundamental parameters. For models with short-range interactions these are the dimensionality of space, and the symmetry of the order parameter. One possibility to determine these scaling relations is to use thermodynamic relations. Here one example is given: The specific heat for constant field c_H and constant magnetization c_M obey the equation

$$\chi_T (c_H - c_M) = T \left(\frac{\partial M}{\partial T}\right)_H^2 \quad . \tag{5.30}$$

Because c_M must be greater or equal to zero one can write the inequality

$$c_H \geq \frac{T \left(\frac{\partial M}{\partial T}\right)_H^2}{\chi_T} \quad . \tag{5.31}$$

Using the definitions of the critical exponents given above, (5.31) can only be fulfilled if

$$\alpha + 2\beta + \gamma \geq 2 \quad , \tag{5.32}$$

which yields one particular scaling relation. It should be noted once again that these inequalities (others can be determined from the convexity properties of the free energy e.g.) are independent of the underlying models they represent universal relations which will proof extremely useful in classifying our models and their range of validity.

5.10 Neutron Diffraction

Neutrons are fermions and carry a magnetic moment due to their spin $s = 1/2$, but no electric charge. They thus not only interact with the nuclei but also with the spins of the crystal electrons, although not with their charge density (as X-rays do). From elastic neutron scattering experiments one can thus obtain information not only about the crystallographic [60, 61] but also about the magnetic lattice. Performing inelastic scattering experiments, the energy and momentum transfer neutrons can create excitations of the spin system such as magnons [62]. From these experiments one can measure e.g. a dispersion relation as described by the Heisenberg Hamiltonian (see Chap. 7) which reads

$$\hbar\omega = D\boldsymbol{k}^2 \quad . \tag{5.33}$$

The so called spinwave stiffness constant D can be determined from inelastic neutron scattering by measuring the excitation of these collective modes via the momentum transfer from the neutron.

5.11 Further Experimental Methods

From the thermodynamical relations there are a number of experiments which are related to the magnetic properties such as specific heat, thermal expansion, elastic constants etc. A further group of investigations are the resonance experiments such as *Mössbauer, NMR, NQR* [63, 64] and the *deHaas–van Alphen* effect which provides a means to measure the dimensions of the Fermi surface [70, 71].

Spectroscopical methods such as *photoemission* [72, 73] allows one to determine the density of states at the Fermi energy, although they have the shortcoming that they are very surface sensitive (which can also be an advantage).

A very exciting method has been developed by using polarized X-rays from a synchrotron namely the *Spin Resolved X-ray Dichroism*, where the differing absorption of a right-circular and a left-circular X-ray beam can be related to the number of spin-up and spin-down electrons in the sample [65].

Spin-resolved scanning tunneling microscopy [67]–[68] can be performed by using a CrO_2 tip, which is a half metallic ferromagnet [66].

6. Weiss Molecular Field Model

To explain the spontaneous magnetic order in solids, Pierre Weiss [16] postulated the existence of an internal magnetic field, the so called molecular field, which should be responsible for the spontaneous magnetic order. About the physical interpretation of that field nothing was known.

One writes the molecular (magnetic) field as

$$H_M = NM \quad , \tag{6.1}$$

where M is the magnetization and N a proportionality factor called molecular field constant. For the derivation one now follows classical thermodynamics. The magnetic moment of a particle with quantum numbers J and m_J is given by:

$$\mu = m_J g_j \mu_B, \qquad \text{for} \quad -J \le m \le +J \quad , \tag{6.2}$$

where m_J and J are the usual quantum numbers and g_j is the Landé factor (5.15). The magnetic moment is measured in the direction of m_J which is the eigenvalue of the J_z component of the $\boldsymbol{J}$ operator. For a given value of $J = 2$, Fig. 6.1 shows the allowed directions of J. The length of the angular momentum vector is $\sqrt{J(J+1)}$ being the square-root of the expectation value of $\boldsymbol{J}^2$.

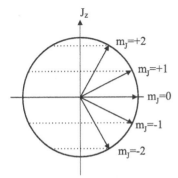

Fig. 6.1. Possible orientations of the angular momentum according to the allowed values of m_J

The magnetic moment M at temperature T is given by the statistical average over all possible states of m_J

$$\frac{M}{M_0} = \frac{1}{Jg_j\mu_B}\frac{\sum\limits_{m_J=-J}^{+J} m_J g_j \mu_B \exp\left(-\frac{W}{k_BT}\right)}{\sum\limits_{m=-J}^{+J} \exp\left(-\frac{W}{k_BT}\right)} , \tag{6.3}$$

with

$$W = -m_J g_j \mu_B H, \quad H = H_M + H_{\text{ext}} . \tag{6.4}$$

M_0 is the magnetic moment at $T = 0$, H_{ext} is an externally applied field. In the Weiss model the magnetic moment is given by an average over a thermally induced statistical distribution of the possible direction of J. Figure 6.2 sketches this distribution the case $J = 2$ and a temperature below T_c.

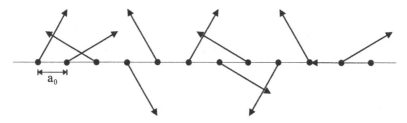

Fig. 6.2. Statistical distribution of the local moments in the Weiss model ($T < T_c$). a_0 is the lattice parameter

To calculate the expression in (6.3) one introduces the following abbreviations

$$a = \frac{Jg_j\mu_B H}{k_BT} , \quad x = \exp\left(\frac{a}{J}\right) = \exp\left(\frac{g_j\mu_B H}{k_BT}\right) \tag{6.5}$$

$$\Rightarrow \quad \frac{M}{M_0} = J^{-1}\frac{\sum_{-J}^{+J} m_J x^{m_J}}{\sum_{-J}^{+J} x^{m_J}} . \tag{6.6}$$

To evaluate the expression given in (6.6) one applies the identity

$$\frac{\sum m_J x^{m_J}}{\sum x^{m_J}} = x\frac{\frac{d}{dx}\sum x^{m_J}}{\sum x^{m_J}} = x\frac{d}{dx}\ln\sum x^{m_J} . \tag{6.7}$$

$$\sum_{-J}^{+J} x^{m_J} = \frac{x^{-J}\left(1 - x^{2J+1}\right)}{1 - x} = \frac{x^{-J} - x^{J+1}}{1 - x} ,$$

$$\ln\sum x^{m_J} = \ln\left(x^{-J} - x^{J+1}\right) - \ln\left(1 - x\right) ,$$

$$x \frac{d}{dx} \ln \sum x^{m_J} = \frac{-Jx^{-J} - (J+1)x^{J+1}}{x^{-J} - x^{J+1}} + \frac{x}{1-x}$$

$$= \frac{-J\left(x^{-J} + x^{J+1}\right)}{x^{-J} - x^{J+1}} - \frac{x^{J+1}}{x^{-J} - x^{J+1}} + \frac{x}{1-x}$$

$$= \frac{J\left(x^{2J+1} + 1\right)}{\left(x^{2J+1} - 1\right)} - \frac{x^{J+1}}{x^{-J} - x^{J+1}} - \frac{x}{x-1}$$

$$= \frac{J\left(x^{2J+1} + 1\right)}{\left(x^{2J+1} - 1\right)}$$

$$+ \frac{1}{2} \left(\frac{x^{J+1} + x^{-J} + x^{J+1} - x^J}{x^{J+1} - x^{-J}} - \frac{x+1+x-1}{x-1} \right)$$

$$= \left(J + \frac{1}{2}\right) \frac{x^{2J+1} + 1}{x^{2J+1} - 1} - \frac{1}{2}\left(\frac{x+1}{x-1}\right) \quad .$$

resubstituting the abbreviations the final result becomes

$$\frac{M}{M_0} = B(a, J) = \frac{2J+1}{2J} \coth\left(a \frac{2J+1}{2J}\right) - \frac{1}{2J} \coth\left(\frac{a}{2J}\right) \quad , \quad (6.8)$$

where $B(a, J)$ is the Brillouin function. The J dependence $B(a, J)$ is shown in Fig. 6.3.

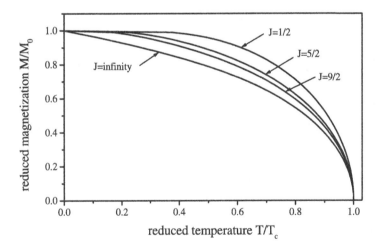

Fig. 6.3. J dependence of the Brillouin function

How perfectly this model works for paramagnetic ions [74] can be seen from the experimental results given in Fig. 6.4.

For high temperature $a \ll 1$ and one can expand the coth according to

$$\coth x = \frac{1}{x} + \frac{x}{3} - \dots \quad , \quad (6.9)$$

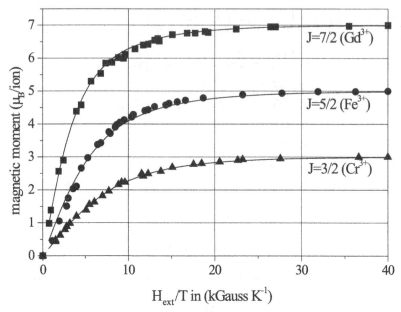

Fig. 6.4. Plot of the magnetic moment versus H_{ext}/T for the metal ions Cr^{3+}, Fe^{3+} and Gd^{3+}. After W.E. Henry [74]

to obtain a much simpler form for the Brillouin function

$$B(a, J) \cong \frac{1}{3} a \frac{J+1}{J} \quad . \tag{6.10}$$

With this relation one can immediately derive an expression for the Curie temperature. As M becomes zero at T_c (6.3) also determines $T_c(H_{ext} = 0)$

$$\frac{M}{M_0} = \frac{1}{3} \frac{J+1}{J} \frac{Jg_J\mu_B}{k_B T} NM \quad ,$$

$$\Rightarrow k_B T_c = \frac{1}{3} (J+1) g_J \mu_B N M_0 \quad . \tag{6.11}$$

For electrons it is obvious that $J = S = \frac{1}{2}$ and $|g_j| = 2$, so that (6.11) becomes

$$k_B T_c = \mu_B N M_0 \quad . \tag{6.12}$$

One can use (6.12) to determine the molecular field constant and the molecular field for Fe, Co, and Ni from the experimental data of the Curie temperature and the magnetic moment (Table 6.1)

The calculated fields are found to be extremely high. Later (see Sect. 16.1) the origin for this "field" will be shown to be the exchange interaction. The exchange interaction is of purely quantum mechanical origin and in lowest

Table 6.1. Molecular field constant N and the respective molecular field H_M for Fe, Co, and Ni

	$N\,(\text{T}/\mu_B)$	$H_M\,(\text{T})$
Fe	700	1500
Co	1300	2100
Ni	1600	940

order of Coulomb type. This "electrostatic" type of interaction rather than the assumed "magnetic" one is the reason for the large numbers in terms of a magnetic field.

For a paramagnetic system the molecular field constant is zero. One obtains the susceptibility by differentiating the magnetic moment M induced by H_{ext} with respect to H_{ext} which yields the Langevin–Debye formula (1.1) from a quantum mechanical Ansatz

$$\chi = \frac{1}{3}J(J+1)\frac{g_j^2\mu_B^2}{k_B T} = \frac{\mu_J^2}{3k_B T} \quad , \tag{6.13}$$

where μ_J is the quantum mechanical value of the magnetic moment due to the angular momentum

$$\mu_J^2 = J(J+1)g_j^2\mu_B^2 = \mu_{\text{eff}}^2 \quad . \tag{6.14}$$

In general (6.14) defines an effective moment μ_{eff} which experimentally is derived from measurements of the Curie constant. As can be seen from the definition, the effective moment measures the size of the local moments present above the ordering temperature. It should be noted that this interpretation via local moments is only valid in systems where these local moments actually determine the magnetic behaviour. For itinerant electron magnetism (Chap. 12) and for spin fluctuations (Chap. 18) this interpretation must be seen in a wider context of moments existing locally in space and time.

Equation (6.11) also allows one to calculate the susceptibility above T_c

$$\frac{M}{M_0} = \frac{1}{3}\frac{J+1}{J}\frac{Jg_j\mu_B}{k_B T}(H_{\text{ext}} + NM)$$

$$= \frac{1}{3}\frac{J+1}{J}\frac{Jg_j\mu_B}{k_B T}\left(H_{\text{ext}} + \frac{3k_B T_c M}{(J+1)g_j\mu_B M_0}\right) \quad ,$$

$$\Rightarrow \frac{M}{M_0}\left(1 - \frac{T_c}{T}\right) = \frac{1}{3}\frac{(J+1)g_j\mu_B}{k_B T}H_{\text{ext}} \quad ,$$

$$\Rightarrow \chi = \frac{M}{H_{\text{ext}}} = \frac{C}{T - T_c} \quad , \quad \text{with} \quad C = \frac{1}{3}(J+1)\frac{g_j\mu_B}{k_B}M_0 \tag{6.15}$$

With the definitions for M_0 (6.3) and μ_{eff}^2 (6.14) one easily recovers the general expression for the Curie constant

$$C = \frac{\mu_{\text{eff}}^2}{3k_{\text{B}}} \quad . \tag{6.16}$$

For $J = S = \frac{1}{2}$ and $|g_j| = 2$, (6.15) reduces to: $C = \frac{\mu_{\text{B}} M_0}{k_{\text{B}}}$. The relation given by (6.15) is the Curie–Weiss law. For high enough temperature all systems obey this linear temperature dependence of the inverse susceptibility as it represents the classical limit.

The spontaneous magnetization as a function of temperature is also given by the Brillouin function. Using (6.1) and (6.11) one gets

$$\frac{M}{M_0} = B\left(\frac{3J}{J+1}\frac{M}{M_0}\frac{T_{\text{c}}}{T}, J\right) \quad , \tag{6.17}$$

which again for $J = S = \frac{1}{2}$ and $|g_j| = 2$ gives

$$\frac{M}{M_0} = \tanh\left(\frac{M}{M_0}\frac{T_{\text{c}}}{T}\right) \quad . \tag{6.18}$$

In the case for $J = \infty$ one arrives at the classical limit where the tilting angle is continuous. The Brillouin function (6.8) reduces to an expression which is usually referred to as the Langevin function[12] which reads

$$\frac{M}{M_0} = \coth(a) - \frac{1}{a} \quad . \tag{6.19}$$

Equation (6.18) is a transcendental equation which is usually solved graphically or numerically. However, approximate solutions can be found around $T = 0$ and $T = T_{\text{c}}$. Expanding (6.18) around T_{c} one immediately obtains the mean field exponent $\beta = \frac{1}{2}$

$$\frac{M}{M_0} = \left(3\left(\frac{T_{\text{c}}}{T} - 1\right)\right)^{\frac{1}{2}} \simeq \left(\frac{T_{\text{c}}}{T} - 1\right)^{\beta} \quad . \tag{6.20}$$

Expanding (6.18) around $T = 0$ gives

$$\frac{M}{M_0} = 1 - 2\exp\left(-\frac{2T_{\text{c}}}{T}\right) \quad . \tag{6.21}$$

Equation (6.18) describes a universal relation between the magnetization M and the temperature T. Unfortunately this relation does not really apply to metallic systems and alloys. As being typical for all mean field models, the Weiss model predicts critical exponents $\beta = 1/2$ and $\gamma = 1$. These exponents deviate from experiment which gives $\beta \approx 0.33$ and $\gamma \approx 1.3 - 1.35$. Only for systems where one finds only a weak interaction between the individual spins, does the Weiss model work satisfactorily.

Five final remarks:

1. The Weiss molecular field model is a mean field model where the interaction between the spins is assumed to have the form of a uniform

field. Quantum mechanical principles enter via the assumption of discrete energy levels associated with the quantum number m_J. The thermal excitation of these localized states can be calculated via classical thermodynamics. These assumptions lead to a restricted applicability of the theory. The Weiss model can be applied to systems with localized (→Hund's rules !) magnetic moments and at high temperatures where the Weiss model becomes classical.

2. The temperature dependence of the magnetic moment is much more often given by a power law rather than the complicated relation in (6.18).
3. The Curie–Weiss behavior is hardly ever found in metallic systems (only at very high temperatures).
4. The Weiss model can easily be applied over limited temperature ranges where it is valid to use the given approximations.
5. At the Curie point, the Weiss model predicts that for temperatures $T \geq T_{\mathrm{c}}$ the spin order vanishes completely, which is mathematically expressed by the finding that the Curie–Weiss behavior of the susceptibility sets in exactly at T_{c} which is called the "paramagnetic Curie temperature". In reality there exist short range magnetic correlations also above T_{c} which should be treated on the basis of quantum statistics [75] which also would lead to a lower Curie temperature which is called "ferromagnetic Curie temperature".

6.1 Rhodes–Wohlfarth Plot

According to the assumption made in the beginning, the Weiss model is a model for weakly interacting localized magnetic moments. This also means that deviations from this behavior can be used for the classification of localized versus itinerant magnetism. If one assumes that at $T = 0K$ the J levels are occupied according to Hund's rule, the magnetic moment per atom is given by

$$\frac{M_0}{n\mu_{\mathrm{B}}} = q_{\mathrm{s}} = g_j J = 2J \quad , \tag{6.22}$$

where M is the magnetic moment of the system, n is the total number of atoms, and $|g_j| = 2$ if the carriers of the magnetic moments are assumed to be electrons. The quantity q_{s} is the "magnetic carrier" per atom.

Calculating the Curie constant one obtains

$$\begin{aligned}
C &= \frac{1}{3}(J+1)g_j\mu_{\mathrm{B}}\frac{n^2 g_j \mu_{\mathrm{B}} J}{nk_{\mathrm{B}}} \\
&= \frac{1}{3}g_j J(g_j J + g_j)\frac{n^2 \mu_{\mathrm{B}}^2}{nk_{\mathrm{B}}} \\
&= \frac{1}{3}q_{\mathrm{c}}(q_{\mathrm{c}}+2)\frac{n^2 \mu_{\mathrm{B}}^2}{R} \quad ,
\end{aligned} \tag{6.23}$$

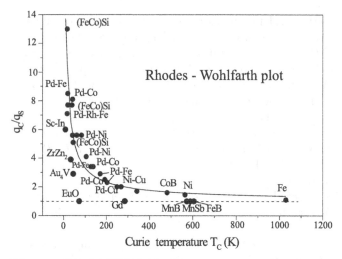

Fig. 6.5. Rhodes–Wohlfarth plot

where R is the gas-constant ($R = Lk_B$) with $L = n$ being Avogadro's number. $q_c = g_j J$ is again defined as a magnetic carrier, but is now calculated from the Curie constant and is thus a measure for the behavior of the system for $T > T_c$. Taking q_s and q_c from experiment and plotting the ratio q_c/q_s as a function of the Curie temperature of the respective system one obtains the Rhodes–Wohlfarth plot [76] (Fig. 6.5)

This phenomenological curve allows one to analyze the different mechanisms of magnetic order. For a system with localized moments, the value of the moment will not change very much if measured below and above T_c. The ratio q_c/q_s will thus be of the order of 1. In the case of delocalized (itinerant) moments $q_s \to 0$, if $T_c \to 0$ while q_c remains independent of T which in turn leads to a large value for q_c/q_s. One thus has two basically different mechanisms:

1. Local magnetic moments:
   This means that the magnetic moments are made up according to Hund's rule. The driving interaction is the *intraatomic exchange* so that interaction with the neighboring atoms is small. The magnetic moment is given by

$$\mu_J = g_j \mu_B \frac{J}{\hbar} \quad , \tag{6.24}$$

   where J is the total angular momentum.
2. Itinerant magnetic moments:
   The carriers of the magnetic moment are the delocalized (itinerant) valence electrons. The driving interaction is the *interatomic exchange* (the exchange interaction of the electron gas). Magnetic order can be descri-

bed as a long-range, coulomb-type interaction. The magnetic moments are not given by the angular momentum.

The Rhodes–Wohlfarth plot shows a fairly uniform distribution of the various systems. This means that cases of complete localization or complete delocalization are hardly ever found. In fact both phenomena exist side by side to a greater or lesser degree. A "unified" theory of solid state magnetism will have to interpolate between these two extremes.

7. Heisenberg Model

In 1928 Werner Heisenberg [22, 23] formulated a model to describe the interaction between neighboring spins which leads to long range ferromagnetic order. Unlike the Weiss model, where the interaction is put into a "mean field", the Heisenberg model accounts in a microscopic way for the pairwise interaction between spins on different lattice sites. In addition the Heisenberg model treats the spins as quantum mechanical observables, rather than the Weiss model where "classical" moments are assumed. In its simple form the Heisenberg Hamiltonian $\mathcal{H}$ [77] reads

$$\mathcal{H} = -I_\mathrm{h} \sum_{l,\delta} S_l S_{l+\delta} - g_j \mu_\mathrm{B} H_\mathrm{ext} \sum_l S_{zl} \quad . \tag{7.1}$$

One assumes that the spins sit on lattice sites l and δ is taken symbolically to replace the vector pointing to the nearest-neighbor sites. I_h is the so called exchange integral which is defined to be positive for the ferromagnetic and negative for the antiferromagnetic case. H_ext is an external field, which is applied along the z axis. This field is > 0 so that at $T = 0K$ all spins are aligned parallel to this field and form the ground state which is described by a state vector $|0\rangle$. (for simplicity $\hbar$ is set to 1). If one considers an ensemble of N atoms with spin S the ground state vector in the $|S, M_S\rangle$ basis reads $|0\rangle \equiv |NS, NS\rangle$. For the total spin the usual relations hold:

$$\mathcal{S}^2 |0\rangle = NS(NS+1) |0\rangle \quad \text{with} \quad \mathcal{S} = \sum_l S_l \quad , \tag{7.2}$$

$$\mathcal{S}_z |0\rangle = NS |0\rangle \quad \text{with} \quad \mathcal{S}_z = \sum_l S_{zl} \quad . \tag{7.3}$$

It will be shown that the spectrum of the Heisenberg Hamiltonian describes excitations of the whole spin system. These excitations are called spin waves or magnons (the term magnon is used in analogy to the term phonon and denotes the respective quantized quasiparticle).

7.1 Magnon Operators

The transformation from the components of the spin vector operator to the creation and annihilation operators for the magnons was devised by Holstein and Primakoff [78] who in 1940 formulated boson operators a^+, a which are used to represent the spin operators. For the components of the spin operator the commutation relation $[S_x, S_y] = iS_z$ (and cyclic permutations) has to be obeyed. This is the reason why a new set of operators is introduced namely

$$S^+ = S_x + iS_y \quad , \quad S^- = S_x - iS_y \quad , \quad \eta = S - S_z \quad , \tag{7.4}$$

where η is a new operator which describes the difference between the total spin and its z component. For the excitation which one will find η will be the respective order parameter since the energy difference $\eta |S, M_S\rangle$ is exactly the energy which is absorbed by the spinwave. If the eigenvalue of η is n the respective eigenfunction of the spinwave can be written as ψ_n. In general the creation and annihilation operators are defined as

$$a^+\psi_n = \sqrt{n+1}\psi_{n+1} \quad , \quad a\psi_n = \sqrt{n}\psi_{n-1} \quad , \tag{7.5}$$

with the boson commutation rule

$$\left[a, a^+\right] = \mathbf{1} \quad , \tag{7.6}$$

and the property

$$\langle\psi_n| a^+a |\psi_n\rangle = n \quad . \tag{7.7}$$

Applying the operators S^+ and S^- on the ground state yields

$$S^+ |S, M_S\rangle = \sqrt{(S - M_S)(S + M_S + 1)} |S, M_S + 1\rangle \quad , \tag{7.8}$$

$$S^- |S, M_S\rangle = \sqrt{(S + M_S)(S - M_S + 1)} |S, M_S - 1\rangle \quad , \tag{7.9}$$

or using the wavefunction ψ_n

$$S^+\psi_n = \sqrt{(S - M_S)(S + M_S + 1)}\psi_{n-1} \quad ,$$

$$S^-\psi_n = \sqrt{(S + M_S)(S - M_S + 1)}\psi_{n+1} \quad ,$$

so that S^+, S^- and η can be formulated in the a, a^+ operators

$$S_l^+ = \sqrt{2S}\left(1 - \frac{1}{2S}a_l^+a_l\right)^{\frac{1}{2}} a_l \quad ,$$

$$S_l^- = \sqrt{2S}a_l^+\left(1 - \frac{1}{2S}a_l^+a_l\right)^{\frac{1}{2}} \quad , \tag{7.10}$$

$$\eta = a_l^+a_l \quad , \quad S_{zl} = S_l - a_l^+a_l \quad .$$

The next step is to transform the operators S^+, S^-, η from the (real space) spin operators a_l^+, a_l to the (reciprocal space) magnon operators b_k, b_k^+, which are related via a lattice Fourier transform given by

$$b_k \equiv \frac{1}{\sqrt{N}}\sum_l \exp\left(i\mathbf{k}\mathbf{x}_l\right) a_l \quad , \quad b_k^+ \equiv \frac{1}{\sqrt{N}}\sum_l \exp\left(-i\mathbf{k}\mathbf{x}_l\right) a_l^+ \quad , \tag{7.11}$$

$$a_l \equiv \frac{1}{\sqrt{N}}\sum_k \exp\left(-i\mathbf{k}\mathbf{x}_l\right) b_k \quad , \quad a_l^+ \equiv \frac{1}{\sqrt{N}}\sum_k \exp\left(i\mathbf{k}\mathbf{x}_l\right) b_k^+ \quad . \tag{7.12}$$

again with the boson commutation rules $\left[b_{k_1}, b_{k_2}^+\right] = \delta_{k_1,k_2}$ and $\left[b_{k_1}, b_{k_2}\right] = \left[b_{k_1}^+, b_{k_2}^+\right] = 0$. The operator b_k^+ creates a magnon with the wave vector $\boldsymbol{k}$, whereas b_k annihilates it. The discrete values for $\boldsymbol{k}$ are defined by the periodic boundary conditions. From the backtransformation (7.12) one notices immediately that the change of an individual state at site l (e.g. a spin flip in a spin $\frac{1}{2}$ systems) is described via a superposition of an infinite number of lattice periodic spin waves which have to be added up at site l to describe this very spin flip.

To perform the transformation to the magnon operators, one has to restrict the description to low excited states only. In this case the square root in (7.10) can be expanded according to $\sqrt{1-\xi} \simeq 1 - \frac{\xi}{2} + \dots$ given that the excitation described by $a_l^+ a_l$ remains small as compared to the total spin $2S$. One obtains

$$S_l^+ = \sqrt{\frac{2S}{N}}\left[\sum_k \exp\left(-i\boldsymbol{k}\boldsymbol{x}_l\right) b_k\right.$$
$$\left. - \frac{1}{4SN} \sum_{k_1 k_2 k_3} \exp\left(i\boldsymbol{x}_l\left(\boldsymbol{k}_1 - \boldsymbol{k}_2 - \boldsymbol{k}_3\right)\right) b_{k_1}^+ b_{k_2} b_{k_3} + \dots\right],$$

$$S_l^- = \sqrt{\frac{2S}{N}}\left[\sum_k \exp\left(i\boldsymbol{k}\boldsymbol{x}_l\right) b_k^+\right.$$
$$\left. - \frac{1}{4SN} \sum_{k_1 k_2 k_3} \exp\left(i\boldsymbol{x}_l\left(\boldsymbol{k}_1 + \boldsymbol{k}_2 - \boldsymbol{k}_3\right)\right) b_{k_1}^+ b_{k_2}^+ b_{k_3} + \dots\right],$$

$$S_{zl} = S - \frac{1}{N}\sum_{k_1 k_2} \exp\left(i\boldsymbol{x}_l\left(\boldsymbol{k}_1 - \boldsymbol{k}_2\right)\right) b_{k_1}^+ b_{k_2} \quad,$$

$$\mathcal{S}_z = NS - \frac{1}{N}\sum_{l k_1 k_2} \exp\left(i\boldsymbol{x}_l\left(\boldsymbol{k}_1 - \boldsymbol{k}_2\right)\right) b_{k_1}^+ b_{k_2} \tag{7.13}$$

$$= NS - \sum_{k_1 k_2} \delta_{k_1 k_2} b_{k_1}^+ b_{k_2} = NS - \sum_k b_k^+ b_k \quad,$$

where $\mathcal{S}_z$ is again the operator for the total spin as defined in (7.3). From the expression for $\mathcal{S}_z$ one notices that $b_k^+ b_k$ can be seen as the occupation number operator for the magnon state $\boldsymbol{k}$, the eigenvalues of $b_k^+ b_k$ are the positive integers.

7.2 Heisenberg Hamiltonian in Magnon Variables

To formulate the Hamiltonian given by (7.1) using the magnon creation and annihilation operators one has to rewrite it as follows

$$\mathcal{H} = -I_\mathrm{h} \sum_{l\delta} \left[S_{zl} S_{z(l+\delta)} + \frac{1}{2} \left(S_l^+ S_{l+\delta}^- + S_l^- S_{l+\delta}^+ \right) \right] - g_j \mu_\mathrm{B} H_\mathrm{ext} \sum_l S_{zl} \ .$$

$$(7.14)$$

The four terms appearing in (7.14) yield

(1) $-I_\mathrm{h} \sum_{l\delta} S_{zl} S_{z(l+\delta)}$

$$= -I_\mathrm{h} \sum_{j\delta} \left[S^2 - \frac{S}{N} \sum_{k_1 k_2} \mathrm{e}^{\mathrm{i}\boldsymbol{x}_{l+\delta}(\boldsymbol{k}_1 - \boldsymbol{k}_2)} b_{k_1}^+ b_{k_2} - \frac{S}{N} \sum_{k_1 k_2} \mathrm{e}^{\mathrm{i}\boldsymbol{x}_l(\boldsymbol{k}_1 - \boldsymbol{k}_2)} b_{k_1}^+ b_{k_2} \right.$$
$$\left. - \frac{1}{N^2} I_\mathrm{h} \sum_{l\delta} \sum_{k_1 k_2 k_3 k_4} \mathrm{e}^{\mathrm{i}\boldsymbol{x}_l(\boldsymbol{k}_1 - \boldsymbol{k}_2) + \mathrm{i}\boldsymbol{x}_{l+\delta}(\boldsymbol{k}_3 - \boldsymbol{k}_4)} b_{k_1}^+ b_{k_2} b_{k_3}^+ b_{k_4} \right. \quad , \quad (7.15)$$

(2) $-\frac{I_\mathrm{h}}{2} \sum_{l\delta} S_l^+ S_{l+\delta}^-$

$$= -\frac{I_\mathrm{h} S}{N} \sum_{l\delta} \left\{ \left[\sum_{k_1} \mathrm{e}^{-\mathrm{i}\boldsymbol{x}_l \boldsymbol{k}_1} b_{k_1} - \frac{1}{4SN} \sum_{k_1 k_2 k_3} \mathrm{e}^{\mathrm{i}\boldsymbol{x}_l(\boldsymbol{k}_1 - \boldsymbol{k}_2 - \boldsymbol{k}_3)} b_{k_1}^+ b_{k_2} b_{k_3} \right] \right.$$
$$\left. \times \left[\sum_{k_1} \mathrm{e}^{\mathrm{i}\boldsymbol{x}_{l+\delta} \boldsymbol{k}_1} b_{k_1}^+ - \frac{1}{4SN} \sum_{k_1 k_2 k_3} \mathrm{e}^{\mathrm{i}\boldsymbol{x}_{l+\delta}(\boldsymbol{k}_1 + \boldsymbol{k}_2 - \boldsymbol{k}_3)} b_{k_1}^+ b_{k_2}^+ b_{k_3} \right] \right\}$$

$$(7.16)$$

(3) $-\frac{I_\mathrm{h}}{2} \sum_{l\delta} S_l^- S_{l+\delta}^+$

$$= -\frac{I_\mathrm{h} S}{N} \sum_{j\delta} \left\{ \left[\sum_{k_1} \mathrm{e}^{\mathrm{i}\boldsymbol{x}_l \boldsymbol{k}_1} b_{k_1}^+ - \frac{1}{4SN} \sum_{k_1 k_2 k_3} \mathrm{e}^{\mathrm{i}\boldsymbol{x}_l(\boldsymbol{k}_1 + \boldsymbol{k}_2 - \boldsymbol{k}_3)} b_{k_1}^+ b_{k_2}^+ b_{k_3} \right] \right.$$
$$\left. \times \left[\sum_{k_1} \mathrm{e}^{-\mathrm{i}\boldsymbol{x}_{l+\delta} \boldsymbol{k}_1} b_{k_1} - \frac{1}{4SN} \sum_{k_1 k_2 k_3} \mathrm{e}^{\mathrm{i}\boldsymbol{x}_{l+\delta}(\boldsymbol{k}_1 - \boldsymbol{k}_2 - \boldsymbol{k}_3)} b_{k_1}^+ b_{k_2} b_{k_3} \right] \right\},$$

$$(7.17)$$

(4) $-g_j \mu_\mathrm{B} H_\mathrm{ext} \sum_l S_{zl}$

$$= g_j \mu_\mathrm{B} H_\mathrm{ext} \frac{1}{N} \sum_{l k_1 k_2} \exp\left(\mathrm{i}\boldsymbol{x}_l (\boldsymbol{k}_1 - \boldsymbol{k}_2)\right) b_{k_1}^+ b_{k_2} - g_j \mu_\mathrm{B} H_\mathrm{ext} \sum_l S_{zl} \quad . (7.18)$$

One now separates the Hamiltonian into three parts $\mathcal{H} = \mathcal{H}_1 + \mathcal{H}_2 + \mathrm{const.}$ where $\mathcal{H}_1$ contains only terms which are bilinear in the magnon operators, $\mathcal{H}_2$ contains all terms of higher order (4th and 6th order) and the constant factors. To calculate the latter part one performs the summation over l and δ assuming z nearest-neighbors giving

$$\mathcal{H} = \mathcal{H}_1 + \mathcal{H}_2 - \underbrace{I_\mathrm{h} N S^2 z}_{\text{from (1)}} - \underbrace{g_j \mu_\mathrm{B} H_\mathrm{ext} N S}_{\text{from (4)}} \quad . \tag{7.19}$$

The bilinear terms are collected to

$$\mathcal{H}_1 = -\frac{I_\mathrm{h} S}{N} \sum_{l \delta k_1 k_2} \left\{ \underbrace{\mathrm{e}^{-\mathrm{i} \boldsymbol{x}_l (\boldsymbol{k}_1 - \boldsymbol{k}_2)} \mathrm{e}^{\mathrm{i} \boldsymbol{k}_2 \boldsymbol{\delta}} \, b_{k_1} b_{k_2}^+}_{\text{from (2)}} + \underbrace{\mathrm{e}^{\mathrm{i} \boldsymbol{x}_l (\boldsymbol{k}_1 - \boldsymbol{k}_2)} \mathrm{e}^{-\mathrm{i} \boldsymbol{k}_2 \boldsymbol{\delta}} \, b_{k_1}^+ b_{k_2}}_{\text{from (3)}} \right.$$

$$\left. - \underbrace{\mathrm{e}^{\mathrm{i} \boldsymbol{x}_l (\boldsymbol{k}_1 - \boldsymbol{k}_2)} \, b_{k_1}^+ b_{k_2}}_{\text{from (1)}} - \underbrace{\mathrm{e}^{\mathrm{i} \boldsymbol{x}_{l+\delta} (\boldsymbol{k}_1 - \boldsymbol{k}_2)} \, b_{k_1}^+ b_{k_2}}_{\text{from (1)}} \right\}$$

$$+ \frac{g_j \mu_\mathrm{B} H_\mathrm{ext}}{N} \sum_{l k_1 k_2} \underbrace{\exp \left(\mathrm{i} \boldsymbol{x}_l (\boldsymbol{k}_1 - \boldsymbol{k}_2) \right) b_{k_1}^+ b_{k_2}}_{\text{from (4)}} \quad . \tag{7.20}$$

Introducing a new quantity $\gamma_k \equiv \frac{1}{z} \sum_{\delta} \exp (\mathrm{i} \boldsymbol{k} \boldsymbol{\delta})$ and performing the summation over l yields

$$\mathcal{H}_1 = -I_\mathrm{h} z S \sum_k \left[\gamma_k b_k b_k^+ + \gamma_{-k} b_k^+ b_k - 2 b_k^+ b_k \right] + g_j \mu_\mathrm{B} H_\mathrm{ext} \sum_k b_k^+ b_k \quad . \tag{7.21}$$

If there exists a center of inversion of the lattice, $\gamma_k = \gamma_{-k}$ so that $\sum_k \gamma_k = 0$. Employing the commutator rule $\left[b_k, b_k^+ \right] = \mathbf{1}$ one finally arrives at

$$\mathcal{H}_1 = \sum_k \underbrace{\left[2 I_\mathrm{h} S z \left(1 - \gamma_k \right) + g_j \mu_\mathrm{B} H_\mathrm{ext} \right]}_{\equiv \, \omega_k} b_k^+ b_k = \sum_k \omega_k b_k^+ b_k \quad . \tag{7.22}$$

The term $\mathcal{H}_2$ contains higher orders in the magnon variables and is usually neglected. A discussion of the magnon–magnon interaction which is described by this term has been given by Dyson [79] and Keffer and Loudon [80].

7.3 Magnon Dispersion Relation

Using the results from the preceding section one writes $\mathcal{H}_1$ as follows

$$\mathcal{H}_1 = \sum_k \mathbf{n}_k \omega_k \tag{7.23}$$

where $\mathbf{n}_k$ is the occupation number operator for a state with energy ω_k. Assuming a Bravais lattice with a center of inversion ($\gamma_k = \gamma_{-k}$) allows one to rewrite γ_k

$$\gamma_k = \frac{1}{z} \sum_\delta \exp\left(i k\boldsymbol{\delta}\right)$$

$$= \frac{1}{2z} \sum_\delta \left[\exp\left(i k\boldsymbol{\delta}\right) + \exp\left(-i k\boldsymbol{\delta}\right)\right]$$

$$= \frac{1}{z} \sum_\delta \cosh\left(i k\boldsymbol{\delta}\right) = \frac{1}{z} \sum_\delta \cos\left(k\boldsymbol{\delta}\right) \quad , \tag{7.24}$$

so that the dispersion relation is obtained in the well known form

$$\omega_k = 2 I_h S z \left(1 - \frac{1}{z} \sum_\delta \cos\left(k\boldsymbol{\delta}\right)\right) + g_j \mu_B H_{\text{ext}} \quad . \tag{7.25}$$

Since the derivation was restricted to low excitation energies, the cosine can be expanded as $\cos x = 1 - \frac{x^2}{2} + \ldots$ so that ω_k becomes

$$\omega_k \simeq I_h S \sum_\delta \left(k\boldsymbol{\delta}\right)^2 + g_j \mu_B H_{\text{ext}} \quad . \tag{7.26}$$

For a cubic lattice with lattice constant a_0 this expression is reduced further

$$\omega_k = g_j \mu_B H_{\text{ext}} + 2 I_h S \left(k a_0\right)^2 \tag{7.27}$$

$$= g_j \mu_B H_{\text{ext}} + D k^2 \quad , \tag{7.28}$$

with D being the spin wave stiffness constant which is an experimentally accessible quantity (from inelastic neutron diffraction e.g.). It should be noted that the factor 2 appearing out of the sum $\sum_\delta \left(k\boldsymbol{\delta}\right)^2$ is not obvious but is due to the properties of the direct and reciprocal lattice. A derivation of this factor is given in Sect. H. Figure 7.1 shows a *classical* representation of a spin wave. The thermally induced motion of the spins appears to be correlated (in contrast to the Weiss model; see Fig. 6.2). Such a type of motion is called *collective excitation*.

Equation (7.27) can be used to estimate the effective mass m^* of a magnon (quasiparticle) by equating the kinetic energy of a massive particle and the excitation energy for $H_{\text{ext}} = 0$,

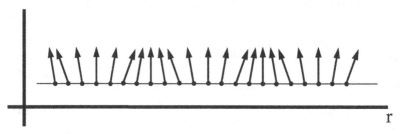

Fig. 7.1. Classical representation of the collective excitation described by the Heisenberg Hamiltonian

$$\frac{k^2}{2m^*} = 2I_\mathrm{h}Sk^2a_0^2$$

$$\Rightarrow m^* = \left(4I_\mathrm{h}Sa_0^2\right)^{-1} \quad . \tag{7.29}$$

For a usual ferromagnet with a Curie temperature around $300K$ one finds an effective mass of about 10 times the electron mass which easily allows one to perform inelastic neutron scattering experiments.

7.3.1 Specific Heat of Magnons

In the long wavelength limit $|k\delta| \ll 1$ and at low temperatures (a few Kelvins) one writes the total energy of the magnon Bose gas as

$$U = \sum_k \omega_k \langle n_\mathrm{B}\rangle \quad , \tag{7.30}$$

with $\langle n_\mathrm{B}\rangle$ being the Bose distribution function

$$\langle n_\mathrm{B}\rangle = \left[\exp\left(\frac{E}{k_\mathrm{B}T}\right) - 1\right]^{-1} \quad . \tag{7.31}$$

One thus gets

$$U = \sum_k \omega_k \left[\exp\left(\frac{\omega_k}{k_\mathrm{B}T}\right) - 1\right]^{-1}$$

$$= \frac{1}{(2\pi)^3} \int_0^{k_\mathrm{max}} Dk^2 \left[\exp\left(\frac{Dk^2}{k_\mathrm{B}T}\right) - 1\right]^{-1} \mathrm{d}^3k$$

$$= \frac{1}{2\pi^2} \int_0^{k_\mathrm{max}} Dk^4 \left[\exp\left(\frac{Dk^2}{k_\mathrm{B}T}\right) - 1\right]^{-1} \mathrm{d}k \quad . \tag{7.32}$$

Using the abbreviations $x = \frac{Dk^2}{\tau}$ and $\tau = k_\mathrm{B}T$ one rewrites the expression given in (7.32) as

$$\frac{\tau^{\frac{5}{2}}}{4\pi^2D^{\frac{3}{2}}} \int_0^{x_\mathrm{max}} \mathrm{d}x\, x^{\frac{3}{2}} \frac{1}{\exp\left(x\right) - 1} \quad .$$

Since the integrand decreases rapidly one can approximate the upper limit by ∞ so that the integral can be evaluated analytically and becomes

$$\Gamma\left(\frac{5}{2}\right)\zeta\left(\frac{5}{2},1\right) = \left(\frac{3\sqrt{\pi}}{4}\right)(1.341) \,,$$

$\Gamma\left(\frac{5}{2}\right)$ is the gamma-function and $\zeta\left(\frac{5}{2},1\right)$ is the Riemann zeta-function giving

$$U \simeq \frac{0.45\tau^{\frac{5}{2}}}{\pi^2 D^{\frac{3}{2}}} \quad ,$$

and for the specific heat at constant volume defined as $c_V = \left(\frac{\partial U}{\partial T}\right)_V$

$$c_V = 0.113 k_B \left(\frac{k_B T}{D}\right)^{\frac{3}{2}} \quad . \tag{7.33}$$

This result means that in a system where spin waves are excited, the magnetic part of the low temperature specific heat should be proportional to $T^{\frac{3}{2}}$. Since this dependence can easily be disentangled from the other contributions to the specific heat (free electrons, phonons, electron-phonon coupling), it provides an experimental possibility to derive a value for the spin wave stiffness constant D.

7.3.2 Ordering Temperature

The ordering temperature (Curie temperature) can be calculated from the number of the reversed spins ([81]). The magnetic moment of the spin system is described by the z component of the total number of spins

$$M_S = g_j \mu_B \mathcal{S}_z = g_j \mu_B \left(NS - \sum_k b_k^+ b_k\right) \quad . \tag{7.34}$$

This allows one to rewrite the temperature dependence of the magnetic moment

$$M_S\,(0) - M_S\,(T) = g_j \mu_B \sum_k \langle n_B \rangle$$

$$= \frac{g_j \mu_B}{2\pi^2} \int_0^{k_{max}} \frac{k^2}{\exp\left(\frac{Dk^2}{k_B T}\right) - 1} dk \quad , \tag{7.35}$$

which with the abbreviations as before gives

$$M_S\,(0) - M_S\,(T) = \frac{g_j \mu_B}{2\pi^2} \left(\frac{T}{D}\right)^{\frac{3}{2}} \int_0^{\infty} x^{\frac{1}{2}} \frac{1}{\exp(x) - 1} dx$$

$$= \frac{g_j \mu_B}{2\pi^2} \left(\frac{T}{D}\right)^{\frac{3}{2}} \Gamma\left(\frac{3}{2}\right) \zeta\left(\frac{3}{2}, 1\right)$$

$$= 0.117 g_j \mu_B \left(\frac{k_B T}{D}\right)^{\frac{3}{2}} \quad . \tag{7.36}$$

Equation (7.36) describes the well known $T^{\frac{3}{2}}$ behavior of the magnetic moment of a system of localized interacting spins (Bloch $T^{\frac{3}{2}}$ law). It is easy to rewrite (7.36) in the usual way

$$M_S\,(T) = M_S\,(0)\left(1 - \left(\frac{T}{T_c}\right)^{\frac{3}{2}}\right) \tag{7.37}$$

with $T_c = \left(\frac{M_S\,(0)}{0.117 g_j \mu_B}\right)^{\frac{2}{3}} \frac{D}{k_B}$ $\tag{7.38}$

In the case of a cubic lattice the spin wave stiffness constant $D = 2I_\mathrm{h}Sa_0^2$. Recalling that $M_S(0) = g_j\mu_\mathrm{B}NS$ and setting $N = 1$ one obtains an expression for T_c which reads

$$k_\mathrm{B}T_\mathrm{c} = 1.26\ S^{\frac{5}{3}}\ I_\mathrm{h}a_0^2\ . \tag{7.39}$$

This result should be compared with the expression which was found from the Weiss model (6.11) which, applied to our case $(J = S)$ becomes

$$k_\mathrm{B}T_\mathrm{c} = \frac{1}{3}S(S+1)N\ . \tag{7.40}$$

Note that N is the molecular field constant as introduced in the Weiss model. The basic structure of both equations is very similar. They essentially differ in the power of S, a behavior which is due to the different treatment which includes quantum fluctuations in the Heisenberg case and a quasi classical treatment in the Weiss case.

Equation (7.39) can be used to calculate the value of the exchange integral I_h. As example it is applied to fcc Gd which represents a local moment system with a total spin $S = \frac{7}{2}$. On the basis of a Curie temperature of 300K and a lattice constant of 5.05Å a value of about 100meVÅ^{-2} is found which is of the right order of magnitude as compared to experiment.

7.4 Approximations for the Heisenberg Model

In a general way the Heisenberg model not only takes into account the interaction between the nearest-neighbor shell but in fact sums over a larger number of neighboring shells

$$\mathcal{H} = -\sum_{i\neq j}I_{ij}\boldsymbol{S}_i\boldsymbol{S}_j \tag{7.41}$$

where I_{ij} are the exchange integrals between spins located on sites i and j. From the Heisenberg model various simpler approaches can be derived.

7.4.1 Ising Model

The simplest and most frequently used spin model is the Ising model [82]. The spin vector operator is assumed to be only one-dimensional and has only two states namely spin up $+1$ ($\uparrow$) and spin down -1 ($\downarrow$). One is thus only using the z-component of the spin vector which can be written as

$$\boldsymbol{S}_i = S_i\boldsymbol{e}_z \tag{7.42}$$

where the spin vector is chosen along the z-axis. The resulting Hamiltonian $\mathcal{H}_\mathrm{I}$ looks very familiar, but it must be noted that the major quantum effects represented by the commutation relations between spins have been removed from the model.

$$\mathcal{H}_I = -\sum_{i \neq j} I_{ij} S_i S_j - g_j \mu_B H_{\text{ext}} \sum_i S_i \tag{7.43}$$

The first term in (7.44) is responsible for the cooperative behavior and the possibility of a phase transition. The second term is again the Zeeman interaction between the spin and an external magnetic field. Since the exchange integrals depend on the extent of the overlap between the wavefunctions which varies about exponentially with distance, one can often restrict the summation over the nearest-neighbor shell.

It is relatively easy to solve the Ising model in one dimension [83] but it took some time before Onsager presented his solution of the two-dimensional problem in a mathematical *tour de force* [84]. The three-dimensional Ising model is still not solved analytically, but there exist numerical solutions based on Monte–Carlo simulations. For the one-dimensional Ising model, the phase transition to the ordered state is always at $T = 0$ which means that for any finite temperature long range order is destroyed completely. In the two-dimensional case a phase transition occurs at a finite temperature and the magnetic moment has the following temperature dependence

$$M(T) = \begin{cases} \left[\left(1 - \left(\sinh \frac{2I}{k_B T} \right) \right)^{-4} \right]^{\beta} & T < T_c \\ 0 & T \geq T_c \end{cases} , \tag{7.44}$$

with $\beta = \frac{1}{8}$. The resulting value for the Curie temperature reads

$$T_c = \frac{2}{\ln \left(1 + \sqrt{2} \right)} \frac{I}{k_B} \simeq 2.269 \frac{I}{k_B} \quad . \tag{7.45}$$

7.4.2 XY Model

The XY model goes one step further by assuming a two dimensional spin vector which describes a spin which can rotate with in the xy-plane. The respective spin vector reads

$$\boldsymbol{S}_i = S_i^x \mathbf{e}_x + S_i^y \mathbf{e}_y \quad , \tag{7.46}$$

and the Hamiltonian becomes

$$\mathcal{H}_{XY} = -\sum_{i \neq j} I_{ij} \left(S_i^x S_j^x + S_i^y S_j^y \right) - g_j \mu_B H_{\text{ext}} \sum_i \left(\boldsymbol{H}_{ext} \boldsymbol{S}_i \right) \quad . \tag{7.47}$$

It has been shown [85] that the XY model only has a traditional second order phase transition for a dimensionality larger than 2, a conjecture which is known as the Mermin–Wagner theorem. The XY model has also attracted great attention as a model for spin glasses, superconductivity and liquid helium.

An interesting aspect of the XY model is the fact that anisotropy, which actually is due to $\boldsymbol{LS}$-coupling, can easily be simulated by introducing an anisotropy parameter λ so that

$$\mathcal{H}_{\mathrm{XY}} = -\sum_{i \neq j} I_{ij} S_i^x S_j^x - (1 - \lambda) \sum_{i \neq j} I_{ij} S_i^y S_j^y - g_j \mu_B H_{\mathrm{ext}} \sum_i \left(\boldsymbol{H}_{ext} \boldsymbol{S}_i \right) .$$

$$(7.48)$$

The value $\lambda = 0$ corresponds to an isotropic XY model and $\lambda = 1$ to a *quasi* Ising model with the x-axis chosen as the Ising axis.

7.4.3 Mean Field Solutions of the Heisenberg Model

Since the product of spin operators is difficult to treat if one goes beyond nearest-neighbor interaction one looks for simplifications of the Heisenberg Hamiltonian. One possible way is to replace the pairwise interaction between the spins by the interaction of one spin with the field exerted by all the neighboring ones. This is achieved by replacing the spin operator by its mean value plus the deviations from it (fluctuations)

$$\boldsymbol{S}_k = \langle \boldsymbol{S}_k \rangle + \underbrace{(\boldsymbol{S}_k - \langle \boldsymbol{S}_k \rangle)}_{\text{fluctuations}} .$$

$$(7.49)$$

This expression, which is still exact, now enters the Heisenberg Hamiltonian

$$\mathcal{H} = -\sum_{i \neq j} I_{ij} \left[\langle \boldsymbol{S}_i \rangle + (\boldsymbol{S}_i - \langle \boldsymbol{S}_i \rangle) \right] \left[\langle \boldsymbol{S}_j \rangle + (\boldsymbol{S}_j - \langle \boldsymbol{S}_j \rangle) \right] \quad ;$$

$$(7.50)$$

multiplying out and neglecting terms which are second order in the fluctuations yields the mean field representation

$$\mathcal{H}_{\mathrm{MF}} = \sum_{i \neq j} I_{ij} \langle \boldsymbol{S}_i \rangle \langle \boldsymbol{S}_j \rangle - \sum_{i \neq j} I_{ij} \left(\boldsymbol{S}_j \langle \boldsymbol{S}_i \rangle + \boldsymbol{S}_i \langle \boldsymbol{S}_j \rangle \right)$$

$$= \sum_{i \neq j} I_{ij} \langle \boldsymbol{S}_i \rangle \langle \boldsymbol{S}_j \rangle - \sum_i \boldsymbol{S}_i \left[2 \sum_j I_{ij} \langle \boldsymbol{S}_j \rangle \right] .$$

$$(7.51)$$

The first term in (7.51) is a constant, the second term can be interpreted as a spin interacting with a field given by $\left[2 \sum_j I_{ij} \langle \boldsymbol{S}_j \rangle \right]$ which is produced by all neighboring spins. Since this form of interaction is exactly the same as in the Weiss model, the solution to the problem is once again the Brillouin function (6.8). The respective expression for the Curie temperature is thus given by

$$T_c = \frac{2S\,(S + 1)}{3k_B} \frac{1}{N} \sum_{i \neq j} I_{ij} .$$

$$(7.52)$$

For an fcc and a bcc lattice this gives

$$T_c = \frac{2S(S+1)}{3k_B} \frac{1}{78} (12J_1 + 6J_2 + 24J_3 + 12J_4 + 24J_5) \quad \text{fcc}$$

$$T_c = \frac{2S(S+1)}{3k_B} \frac{1}{58} (8J_1 + 6J_2 + 12J_3 + 24J_4 + 8J_5) \quad \text{bcc}$$

where the summation is taken up to the 5th neighbor shell. The results for T_c are usually in fair agreement with experiment, although they are systematically too high which is typically of the mean field approximation which suppresses additional quantum fluctuations.

8. Itinerant Electrons at 0 K

While Weiss' molecular field theory assumes localized electrons which occupy m_j-dependent energy levels, Stoner theory of itinerant magnetism describes particles which move freely in the periodic potential of the solid as a more or less free electron gas. Since for these crystal electrons the angular momentum is no longer a good quantum number (the orbital moment is quenched by the crystal field; see Sect. C.), their properties are governed by their momentum described by their wave vector $\boldsymbol{k}$ and their spin. These electron states overlap and form electron bands rather than discrete energy levels (see Chap. 4). The solid state analogue to the energy levels is thus the density of states formed by these bands (2.2). Consequently one finds that the magnetic moments of the transition metal are rational numbers rather than integers (as they are for the rare earths, where the f-electrons are localized). These rational numbers cannot be explained from a successive orbital occupation as described by Hund's rules. A model for the magnetism of metals has to take into account this electronic structure which indeed is responsible for all properties which distinguish metals from other solids. The model of itinerant electron magnetism goes back Stoner who formulated it during the 1930s [30].

The Stoner model is based on the following postulates:

1. The carriers of magnetism are the unsaturated spins in the d-band.
2. Effects of exchange are treated within a molecular field term.
3. One must conform to Fermi statistics.

In analogy with the Weiss model one introduces a molecular field which should contain all interactions. The molecular field per atom is given by

$$H_\mathrm{M} = NM = NM_0\zeta \quad , \quad \text{with} \quad \zeta = \frac{M}{M_0} \quad . \tag{8.1}$$

The energy shift exerted by the molecular field is

$$\epsilon_m = -\mu_\mathrm{B} H_\mathrm{M} = -\mu_\mathrm{B} NM_0\zeta = -k_\mathrm{B}\Theta\zeta . \tag{8.2}$$

Equation (8.2) defines a characteristic temperature Θ which is given by

$$\Theta = \frac{\mu_\mathrm{B} NM_0}{k_\mathrm{B}} . \tag{8.3}$$

To calculate the free energy at $T = 0K$ one starts from $H = \mathrm{d}F/\mathrm{d}M$. The contribution of a magnetic field to the free energy is given by

$$\frac{E_{\mathrm{m}}}{n} = -\int H \mathrm{d}M$$

$$= -\int_0^\zeta k_{\mathrm{B}}\Theta\zeta\mathrm{d}\zeta = -\frac{1}{2}k_{\mathrm{B}}\Theta\zeta^2$$

$$\Rightarrow E_{\mathrm{m}} = -\frac{1}{2}nk_{\mathrm{B}}\Theta\zeta^2 \quad , \tag{8.4}$$

where n is the number of particles (electrons).

One starts from a paramagnetic density of states and splits it into two identical bands for spin-up and spin-down (Fig. 8.1).

If one applies an external magnetic field (molecular field) the bands become shifted relative to each other (Fig. 8.2a). Since the Fermi energy has to be the same for both spin directions ($\Rightarrow$ equal chemical potentials) this spin splitting causes a redistribution of electrons and a mutual shift of the two subbands (Fig. 8.2b) which leads to a difference in the occupation numbers for spin-up and spin-down.

The occupation numbers n^+, n^- and the relative magnetization ζ are related to one another by the following:

$$n = n^+ + n^- \quad ,$$
$$n\zeta = n^+ - n^- \quad ,$$
$$n^\pm = \frac{n}{2}(1 \pm \zeta) \quad . \tag{8.5}$$

These occupation numbers are related to the density of states $\mathcal{N}(\varepsilon)$ via

$$\frac{n}{2} = \int_0^{\varepsilon_{\mathrm{F}}} \mathcal{N}(\varepsilon)\,\mathrm{d}\varepsilon \quad ,$$

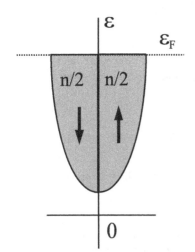

Fig. 8.1. Non-magnetic density of states split into the two subbands for the two spin directions

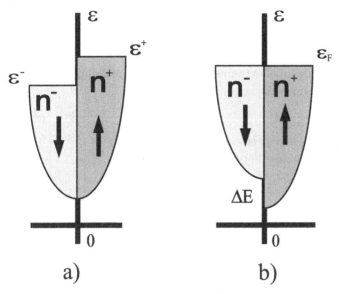

a)   b)

Fig. 8.2. Spin split density of states. n^+, n^- number of electrons for spin up and spin down, $\varepsilon^+, \varepsilon^-$ Fermi energies for spin up and spin down

$$\frac{n}{2}\left(1 \pm \zeta\right) = \int\limits_{0}^{\varepsilon^{\pm}} \mathcal{N}\left(\varepsilon\right) \mathrm{d}\varepsilon \quad ,$$

$$\frac{n}{2}\zeta = \int\limits_{\varepsilon_{\mathrm{F}}}^{\varepsilon^{+}} \mathcal{N}\left(\varepsilon\right) \mathrm{d}\varepsilon \quad , \quad \frac{n}{2}\zeta = \int\limits_{\varepsilon^{-}}^{\varepsilon_{\mathrm{F}}} \mathcal{N}\left(\varepsilon\right) \mathrm{d}\varepsilon \quad . \tag{8.6}$$

If one considers a parabolic band (2.24), where the density of states is proportional to $\sqrt{\varepsilon}$ the spin splitting is then given by

$$\varepsilon^{\pm} = \varepsilon_{\mathrm{F}}\left(1 \pm \zeta\right)^{\frac{2}{3}} \quad . \tag{8.7}$$

Calculating the respective band energy E_{b} by integrating over the occupied electron states gives

$$E_{\mathrm{b}} = \int\limits_{\varepsilon_{\mathrm{F}}}^{\varepsilon^{+}} \varepsilon \mathcal{N}\left(\varepsilon\right) \mathrm{d}\varepsilon - \int\limits_{\varepsilon^{-}}^{\varepsilon_{\mathrm{F}}} \varepsilon \mathcal{N}\left(\varepsilon\right) \mathrm{d}\varepsilon + \text{ const.} \tag{8.8}$$

For a parabolic band $\varepsilon^{\pm}$ are given by (8.6) and (8.7) and one thus obtains for the band energy

$$E_{\mathrm{b}} = \frac{3}{10}n\varepsilon_{\mathrm{F}}\left[(1 + \zeta)^{\frac{5}{3}} + (1 - \zeta)^{\frac{5}{3}}\right] + \text{ const.} \tag{8.9}$$

The free energy is now

$$E = E_b + E_m = E(\zeta)$$
$$= \frac{3}{10}n\varepsilon_F\left[(1+\zeta)^{\frac{5}{3}} + (1-\zeta)^{\frac{5}{3}}\right] - \frac{1}{2}nk_B\Theta\zeta^2 + \text{const.} \tag{8.10}$$

One now determines a possible extremum (minimum or maximum) of $E(\zeta)$, where $\frac{dE(\zeta)}{d\zeta} = 0$ and obtains the condition

$$\varepsilon^+ - \varepsilon^- = 2k_B\Theta\zeta = \Delta E \quad , \tag{8.11}$$

where ΔE is called *molecular field energy* or *band splitting*.

Again for the parabolic band one finds

$$\frac{k_B\Theta}{\varepsilon_F} = \frac{1}{2\zeta}\left[(1+\zeta)^{\frac{2}{3}} - (1-\zeta)^{\frac{2}{3}}\right] \quad . \tag{8.12}$$

Equation (8.12) gives the equilibrium state as a function of ζ:

$$\zeta = 0 \quad \Rightarrow \quad \frac{k_B\Theta}{\varepsilon_F} = \frac{2}{3} \simeq 0.67 \quad ,$$
$$\zeta = 1 \quad \Rightarrow \quad \frac{k_B\Theta}{\varepsilon_F} \geq \frac{1}{\sqrt[3]{2}} \simeq 0.79 \quad . \tag{8.13}$$

This result describes three ranges of magnetic order as shown in Fig. 8.3.

1. There exists a threshold below $\frac{k_B\Theta}{\varepsilon_F} < \frac{2}{3}$ such that no magnetic ordering occurs $\rightarrow$ the system is non-magnetic.
2. Between $\frac{2}{3} < \frac{k_B\Theta}{\varepsilon_F} < 1/\sqrt[3]{2}$ the molecular field is not strong enough to saturate the spins for the majority spin direction $\rightarrow$ the system is weakly ferromagnetic.

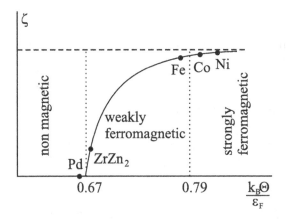

Fig. 8.3. Ranges of magnetic order described within the Stoner model for a parabolic band. Although transition metals do not have parabolic bands their hypothetical positions in the three regimes are given. Pd is at the verge of being magnetic, ZrZn$_2$ is a very weak ferromagnet, Fe is on the verge to the strongly ferromagnetic regime, both Co and Ni are strong ferromagnets

3. For $\frac{k_B \Theta}{\varepsilon_0} \geq 1/\sqrt[3]{2}$ all spins are saturated by the molecular field → the system is strongly ferromagnetic

If one calculates the inverse susceptibility χ^{-1} from (8.8) and (8.10) from the second derivative of $E(\zeta)$ one obtains

$$\frac{n^2 \mu_B^2}{\chi} = \frac{d^2 E(\zeta)}{d\zeta^2} = \frac{n^2}{4} \left(\frac{1}{\mathcal{N}(\varepsilon^+)} + \frac{1}{\mathcal{N}(\varepsilon^-)} \right) - nk_B \Theta \quad . \tag{8.14}$$

Since for a minimum of the free energy the second derivative has to be larger than zero (positive susceptibility), (8.14) provides a criterion for that case. In the non-magnetic limit one has $\zeta = 0$ and obviously $\mathcal{N}(\varepsilon^+) = \mathcal{N}(\varepsilon^-) = \mathcal{N}(\varepsilon_F)$. If the free energy should become smaller for finite ζ, meaning that there is spontaneous magnetic order, $E(\zeta)$ must have a maximum at $\zeta = 0$. This condition leads to the so called Stoner criterion for spontaneous magnetic order of a system of itinerant electrons

$$\frac{2}{n} \mathcal{N}(\varepsilon_F) k_B \Theta \geq 1 \quad . \tag{8.15}$$

The Stoner criterion is fulfilled if either the molecular field term $k_B \Theta$, or the density of states at the Fermi energy $\mathcal{N}(\varepsilon_F)$ is large. It is found that the molecular field constants are of the same order of magnitude for most metallic systems. The (often) continuous phase transition (second order) from the non-magnetic to the ferromagnetic state is thus usually caused by a large value of the DOS at the Fermi energy. It should be noted that these large values of the density of states will never be reached by an electron density which behaves

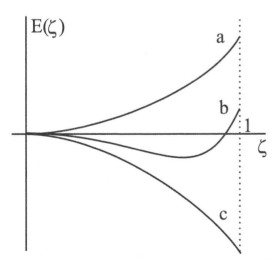

Fig. 8.4. Total energy $E(\zeta)$ as a function of the reduced magnetization ζ for the three magnetic regimes: (**a**) non-magnetic, (**b**) weakly ferromagnetic, (**c**) strongly ferromagnetic

like a free electron gas. Only if the bandwidth is considerably smaller, like for the $3d$-electrons, the DOS is large enough to fulfill the Stoner criterion.

The derivation above follows the original work by E.C. Stoner from 1936. The interpretation of the molecular field as a characteristic temperature is nowadays not felt to have any physical reason. In his original work however, Stoner speculated that this temperature might be somehow related to the Curie temperature. Unfortunately this assumption (of single particle excitations) leads to Curie temperatures which are too high by a factor of $3-5$. It will be shown later that only collective excitations (like those which appear in the Heisenberg model) which depend on the dynamical properties of the susceptibility describe the Curie temperature properly.

In order to formulate the important Stoner criterion in a contemporary way the derivation above is repeated in a more general form which does not depend on the assumption of a parabolic band.

8.1 Pauli Susceptibility of the Itinerant Electrons

Applying an external magnetic field causes a splitting of the spin-up and spin-down Fermi energies according to: $\varepsilon^\pm = \varepsilon_0 \pm \mu_B H_{ext}$. The band energy is then given by

$$E_b = \int_0^{\varepsilon_F} \varepsilon \mathcal{N}(\varepsilon)\,d\varepsilon - \int_{\varepsilon^-}^{\varepsilon_F} \varepsilon \mathcal{N}(\varepsilon)\,d\varepsilon + \int_0^{\varepsilon_F} \varepsilon \mathcal{N}(\varepsilon)\,d\varepsilon + \int_{\varepsilon_F}^{\varepsilon^+} \varepsilon \mathcal{N}(\varepsilon)\,d\varepsilon \quad .(8.16)$$

Carrying out the integration with the assumption of a rectangular DOS ($\mathcal{N}(\varepsilon)$ constant around ε_F) yields

$$
\begin{aligned}
E = E_p &- \frac{\mathcal{N}(\varepsilon_F)}{2}\left(\varepsilon_F^2 - (\varepsilon_F - \mu_B H_{ext})^2\right) \\
&+ \frac{\mathcal{N}(\varepsilon_F)}{2}\left((\varepsilon_F + \mu_B H_{ext})^2 - \varepsilon_F^2\right) \\
&= E_p + \mathcal{N}(\varepsilon_F)\,\mu_B^2 H_{ext}^2 \quad ,
\end{aligned}
\tag{8.17}
$$

where E_p is the contribution of the non spin-split (non-magnetic) density of states. Taking the second derivative of E with respect to H_{ext} yields the Pauli susceptibility of the non-interacting electron gas as before (1.9):

$$\chi_P = 2\mu_B^2 \mathcal{N}(\varepsilon_F) \quad .$$

8.2 Susceptibility of the Interacting Itinerant Electrons

The free energy for the interacting itinerant electrons can be obtained in an analogous way but now by including the molecular field energy $-I_s M^2/2$.

This term is completely analogous to (8.4). I_s is again an effective interaction parameter (like Θ), the so called Stoner exchange integral.

$$E_b = \int_0^{\varepsilon_F} \varepsilon \mathcal{N}(\varepsilon)\,d\varepsilon - \int_{\varepsilon^-}^{\varepsilon_F} \varepsilon \mathcal{N}(\varepsilon)\,d\varepsilon + \int_0^{\varepsilon_F} \varepsilon \mathcal{N}(\varepsilon)\,d\varepsilon + \int_{\varepsilon_F}^{\varepsilon^+} \varepsilon \mathcal{N}(\varepsilon)\,d\varepsilon$$

$$- \frac{I_s M^2}{2} \ . \tag{8.18}$$

Carrying out the same integration procedure as above gives analogously

$$E = E_p - \frac{\mathcal{N}(\varepsilon_F)}{2}\left(\varepsilon_F^2 - (\varepsilon_F - \mu_B H_{ext})^2\right)$$

$$+ \frac{\mathcal{N}(\varepsilon_F)}{2}\left((\varepsilon_F + \mu_B H_{ext})^2 - \varepsilon_F^2\right) - \frac{I_s M^2}{2} \ . \tag{8.19}$$

From the Pauli susceptibility (1.9) one obtains

$$\mu_B^2 H_{ext}^2 \mathcal{N}(\varepsilon_F) = \frac{M^2}{4\mu_B^2 \mathcal{N}(\varepsilon_F)}$$

which allows one to write the free energy as

$$E = E_p + \frac{M^2}{4\mathcal{N}(\varepsilon_F)\mu_B^2} - \frac{I_s M^2}{2} \ , \tag{8.20}$$

which in turn leads again to the susceptibility of the interacting itinerant electrons which now reads

$$\chi = \frac{\chi_P}{1 - 2\mu_B^2 I_s \mathcal{N}(\varepsilon_F)} = \chi_P \, S \ . \tag{8.21}$$

It appears that for the interacting electron system the susceptibility is no longer given by the bare Pauli term, but is enhanced by the factor $S = 1/\left(1 - 2\mu_B^2 I_s \mathcal{N}(\varepsilon_F)\right)$ the Stoner enhancement factor. The discrepancies between experiment and the non-interacting electron gas results can often be explained by the exchange enhancement (see e.g. Pd).

The denominator again allows to formulate the Stoner criterion by demanding that the susceptibility must be positive in the minimum of the free energy. If for $M = 0$, the susceptibility is negative one obtains that

$$2\mu_B^2 I_s \mathcal{N}(\varepsilon_F) > 1 \ . \tag{8.22}$$

The quantity I_s is called Stoner exchange integral (Stoner parameter) and is found to be a quasi-atomic property which depends very little on chemical or metallurgical effects (bonding, alloying etc.). Also the k-vector dependence of I_s is very weak (as one can expect for mean field property). I_s can be calculated from the exchange interaction, since the band splitting is given by the expectation value of the difference of the spin up and spin down exchange potentials. Values derived from electronic band structure calculations [88] employing the KKR-method (Korringa–Kohn–Rostoker method) [89, 90] and the local spin density approximation for exchange and correlation are given in Table 8.1.

Table 8.1. Stoner exchange integral I_s and density of states at the Fermi energy $\mathcal{N}(\varepsilon_F)$ for various metallic elements throughout the periodic table. a_0 is the lattice constant of the respective crystallographic unit cell

Element	Structure	a_0 [bohr]	$\mathcal{N}(\varepsilon_F)$ $\left[\mathrm{Ry}^{-1}\right]$	I_s [mRy]
Li	bcc	6.42	3.25	172
Be	fcc	5.96	0.36	156
Na	bcc	7.7	3.1	134
Mg	fcc	8.4	3.1	104
Al	fcc	7.6	2.8	90
K	bcc	9.45	4.9	98
Ca	fcc	10.0	10.5	74
Sc	bcc	6.74	16.5	50
Sc	fcc	8.49	12	50
Ti	fcc	7.56	11	50
V	bcc	5.54	11	52
Cr	bcc	5.30	4.7	56
Mn	fcc	6.543	10.5	60
Fe	bcc	5.15	21	68
Co	fcc	6.448	15	72
Ni	fcc	6.55	27	74
Cu	fcc	6.76	2	54
Zn	fcc	7.25	2	76
Ga	fcc	7.83	2.8	74
Rb	bcc	10.21	6	86
Sr	fcc	10.88	2.1	62
Y	fcc	9.23	9.5	48
Zr	bcc	6.54	8.5	46
Nb	bcc	6.2	9.5	44
Mo	bcc	5.89	4.5	44
Tc	fcc	7.28	8.5	44
Ru	fcc	7.2	7.5	44
Rh	fcc	7.24	9	48
Pd	fcc	7.42	15.5	50
Ag	fcc	7.79	1.8	60
Cd	fcc	8.40	2.5	64
In	fcc	8.95	3.4	30

8.3 Non-linear Effects

There exist some systems which, although they do not fulfill the Stoner criterion, show a local minimum in the free energy for finite magnetic moment. This behavior eventually leads to an effect called metamagnetism which will be discussed in Sect. 8.4.6. For these systems the application of an external field leads to a first order phase transition from a low-moment to a high-moment state. This means that up to the a critical field H_{c2} the system remains paramagnetic and then jumps to a finite moment. This behavior is caused by a magnetization-dependent term in the susceptibility,

$$\frac{\mathrm{d}^2 E\left(\zeta\right)}{\mathrm{d}\zeta^2} = \frac{n^2}{2\mathcal{N}\left(\varepsilon_F\right)}\left(1 - c\zeta^2\right) - nk_B\Theta \quad . \tag{8.23}$$

with c given by

$$c = \frac{1}{8}\frac{n^2}{\mathcal{N}\left(\varepsilon_F\right)^2}\left[\frac{\mathcal{N}\left(\varepsilon_F\right)''}{\mathcal{N}\left(\varepsilon_F\right)} - 3\left(\frac{\mathcal{N}\left(\varepsilon_F\right)'}{\mathcal{N}\left(\varepsilon_F\right)}\right)^2\right] \quad . \tag{8.24}$$

A detailed derivation of c is given in Sect. E. The equilibrium is again given by $\frac{\mathrm{d}E}{\mathrm{d}\zeta} = 0$

$$k_B\Theta = \frac{n}{2\mathcal{N}\left(\varepsilon_F\right)}\left(1 - \frac{1}{3}c\zeta^2\right) \quad . \tag{8.25}$$

Demanding a minimum at $\zeta = 0$ one finds the condition

$$\frac{2}{n}\mathcal{N}\left(\varepsilon_F\right)k_B\Theta \leq 1 - \frac{1}{3}c\zeta^2 \quad . \tag{8.26}$$

If c is negative, this condition will be fulfilled for all values of ζ. This is for example the case when the Fermi energy lies in a maximum of the density of states where $\mathcal{N}\left(\varepsilon_F\right)' = 0$ and $\mathcal{N}\left(\varepsilon_F\right)'' < 0$. If the Fermi energy lies in a minimum of the DOS, c is positive and the condition will be violated for a finite value of ζ. This means that the system will behave like an ordinary non-magnetic one until this critical value of ζ is reached, where a phase transition to a magnetic state sets in.

For $c < 0$ the magnetization dependence is, however

$$k_B\Theta = \frac{n}{2\mathcal{N}\left(\varepsilon_F\right)}\left(1 + \frac{1}{3}|c|\zeta^2\right) \quad . \tag{8.27}$$

8.4 Effects of High Fields at 0 K

Under an applied magnetic field the equilibrium condition is given by

$$\frac{\mathrm{d}E\left(\zeta\right)}{\mathrm{d}\zeta} = n\mu_B H_{\mathrm{ext}} \quad . \tag{8.28}$$

Since this field adds to the molecular field, it increases the band splitting so that $\varepsilon^{\pm} \to \tilde{\varepsilon}^{\pm}$ and $\zeta \to \tilde{\zeta}$. The band splitting given by (8.11) then reads

$$\tilde{\varepsilon}^{+} - \tilde{\varepsilon}^{-} = 2k_{\mathrm{B}}\Theta\tilde{\zeta} + 2\mu_{\mathrm{B}}H_{\mathrm{ext}} \quad . \tag{8.29}$$

Again one calculates the inverse susceptibility $\chi^{-1} = \frac{\mathrm{d}\tilde{\zeta}}{\mathrm{d}H_{\mathrm{ext}}}$ and obtains the Wohlfarth–Gersdorf [86, 87] form of the high field susceptibility

$$\frac{n\mu_{\mathrm{B}}^{2}}{\chi} = \frac{1}{n}\frac{\mathrm{d}^{2}E\left(\zeta\right)}{\mathrm{d}\zeta^{2}} = \frac{1}{4}n\left(\frac{1}{\mathcal{N}\left(\varepsilon^{+}\right)} + \frac{1}{\mathcal{N}\left(\varepsilon^{-}\right)}\right) - k_{\mathrm{B}}\Theta \quad . \tag{8.30}$$

Equation (8.30) not only describes the susceptibility of a paramagnetic system under an applied field, but of course also the susceptibility of a ferromagnet (8.14) where the bands are split by the "molecular field". From the form of (8.30) it is obvious that the susceptibility becomes small if either $\mathcal{N}\left(\varepsilon^{+}\right)$ and/or $\mathcal{N}\left(\varepsilon^{-}\right)$ are small. This means that strongly ferromagnetic materials, where the spin-up band is filled with electrons and $\mathcal{N}\left(\varepsilon^{+}\right)$ is thus small, usually have a small susceptibility. In weakly itinerant systems where both $\mathcal{N}\left(\varepsilon^{+}\right)$ and $\mathcal{N}\left(\varepsilon^{-}\right)$ are reasonably large also the susceptibility is large as well.

8.4.1 Non-magnetic Limit

In the non-magnetic limit the energies ε^{+} and ε^{-} become equal to ε_{F} so that $\mathcal{N}\left(\varepsilon^{+}\right) = \mathcal{N}\left(\varepsilon^{-}\right) = \mathcal{N}\left(\varepsilon_{\mathrm{F}}\right)$. It is easy to calculate that the high field susceptibility reduces to

$$\begin{aligned} \chi &= 2\mu_{\mathrm{B}}^{2}\mathcal{N}\left(\varepsilon_{\mathrm{F}}\right)S = \chi_{\mathrm{P}}S \\ &= \frac{2\mu_{\mathrm{B}}^{2}\mathcal{N}\left(\varepsilon_{\mathrm{F}}\right)}{1 - \frac{2}{n}k_{\mathrm{B}}\Theta\mu_{\mathrm{B}}^{2}\mathcal{N}\left(\varepsilon_{\mathrm{F}}\right)} \quad . \end{aligned} \tag{8.31}$$

The susceptibility χ_{P} is again the Pauli susceptibility of the non-interacting electron gas, enhanced by the molecular field term. The quantity S is the Stoner enhancement factor. In the case of Pd where the Fermi energy lies in a region of large density of states (3.3), S is of the order $7 - 10$ explaining the earlier mentioned discrepancy between experiment and theory. Since the susceptibility diverges as $S \to \infty$, (8.31) allows one to derive the Stoner criterion as given by (8.15) and (8.22). The effect of exchange enhancement is indispensable for the understanding of metallic magnetism. In the framework of the Stoner model this quantum mechanical many-body effect is described within the mean field approximation leading to an effective exchange parameter I_{s}. Paramagnetic systems with large exchange enhancement are thus on the verge to a ferromagnetic phase transition. One of the most striking examples is the intermetallic compound $TiBe_{2}$ which forms a cubic Laves phase. Although consisting of two normally magnetically rather "inactive" constituents, it exhibits a Stoner enhancement factor of $S \approx 30$. Adding only 0.1% Cu leads to a slight increase of the volume which is sufficient to trigger the phase transition towards magnetic order.

8.4.2 Strong Ferromagnets

Equation (8.30) also describes the susceptibility in the magnetic state when $\varepsilon^+ \neq \varepsilon^-$. In the case of a strong ferromagnet one spin band is fully occupied so that $\mathcal{N}(\varepsilon^+)$ is small or even zero. The susceptibility of such systems is thus also small or zero because the spin splitting is saturated so that an additional field cannot cause an significant additional magnetic moment. hcp Co is a typical example for a strong ferromagnet (Fig. 8.6). The spin-up band is completely filled and the the Fermi energy is separated form the top of the spin-up d-band by the so called *Stoner gap* . The determination of the Stoner gap by means of photo-electron-spectroscopy played an important role for the understanding of magnetic ordering in solids.

8.4.3 Weak Ferromagnets

To calculate the susceptibility of a weak ferromagnet one replaces the two terms appearing in (8.30) by the respective expressions given by (8.24) and (8.25)

$$\frac{n\mu_B^2}{\chi} = \frac{n}{2\mathcal{N}(\varepsilon_F)}\left(1 + |c|\,\zeta^2\right) - \frac{n}{2\mathcal{N}(\varepsilon_F)}\left(1 + \frac{1}{3}|c|\,\zeta^2\right) \quad,$$

$$\Rightarrow \chi = \frac{3\mu_B^2\mathcal{N}(\varepsilon_F)}{|c|\,\zeta^2} \quad, \tag{8.32}$$

where a negative value of c is readily assumed (see discussion above) and ζ is the respective equilibrium moment. Equation (8.32) demonstrates that weak ferromagnets always have a high susceptibility.

8.4.4 bcc Iron and hcp Cobalt

To demonstrate the applicability of the theoretical results the densities of states of representative systems are discussed. As examples of a more weakly ferromagnetic and of a strongly ferromagnetic material Fe and Co are considered, respectively. In the case of bcc Fe (Fig. 8.5) the spin splitting (and the magnetic moment) is determined by the position of the Fermi energy in a minimum of the spin-down DOS. This position is energetically more favorable than for the formation of a strong ferromagnet with a completely filled spin-up band. For both spin directions the value of the DOS at the Fermi energy is finite so that bcc Fe is just weakly ferromagnetic. In Fig. 8.5 also the non-magnetic DOS is shown (dotted line). In the non-magnetic case the density at the Fermi energy is sufficiently large to fulfill the Stoner criterion. Cobalt (Fig. 8.6) has one more electron and a different crystal structure (hcp). Due to the larger number of nearest neighbors the density of states for the closed packed systems (hcp and fcc) is much more compact and does not show these pronounced minima and maxima as the in the bcc modification. Thus hcp

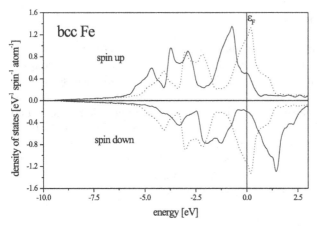

Fig. 8.5. Density of states of bcc Fe. Ferromagnetic state ($M = 2.2\mu_B$) (full line), non-magnetic state (dotted line). In the non-magnetic state, the Fermi energy lies in a region of high density of states which sufficient to fulfill the Stoner criterion

Co is a strong ferromagnet with the majority band being fully occupied. The Fermi energy is positioned above the spin up d-band in a region of very low DOS where mainly s-states exist. The resulting susceptibility is small.

In both systems it is found that the band splitting leaves the shape of the two subbands almost unchanged. This behavior is called *rigid band behavior*. It must be noted that this rigid band picture is at best valid for pure metals or alloys between metals (magnetic) and metalloids (non-magnetic). In alloys which consist of more than one magnetic constituent (e.g. FeCo) the interaction between the different atoms leads to changes in the electronic structure

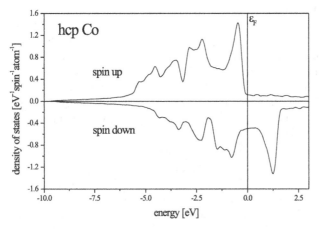

Fig. 8.6. Density of states of ferromagnetic hcp Co ($M = 1.6\mu_B$)

which go beyond this simple picture. The resulting magnetic interaction is called covalent magnetism (Chap. 10).

8.4.5 Extremely High Fields

For extremely high magnetic fields one finds a deviation from the linear relation between field and magnetization. This can be understood by returning to (8.14) and (8.23) to formulate

$$n\mu_B H_{\text{ext}} = \frac{dE}{d\zeta} = \frac{n^2\zeta}{2\mathcal{N}(\varepsilon_F)}\left(1 - \frac{1}{3}c\zeta^2\right) - nk_B\Theta\zeta \quad . \tag{8.33}$$

As a first approximation the ζ^3 term can be neglected

$$n\mu_B H_{\text{ext}} = \frac{n^2\zeta}{2\mathcal{N}(\varepsilon_F)}\frac{1}{S} \quad ,$$

and as a second step ζ is determined from (8.31)

$$n\mu_B H_{\text{ext}} = \frac{n^2\zeta}{2\mathcal{N}(\varepsilon_F)} - nk_B\Theta\zeta - \frac{1}{3}\frac{n^2c}{2\mathcal{N}(\varepsilon_F)}\frac{8\mathcal{N}(\varepsilon_F)^3\mu_B^3}{n^3}S^3 H_{\text{ext}}^3 \quad .$$

Equating both approximations yields

$$\frac{n^2\zeta}{2\mathcal{N}(\varepsilon_F)}\frac{1}{S} = n\mu_B H_{\text{ext}}\left(1 + \frac{1}{6}\left(\frac{\mathcal{N}(\varepsilon_F)''}{\mathcal{N}(\varepsilon_F)} - 3\left(\frac{\mathcal{N}(\varepsilon_F)'}{\mathcal{N}(\varepsilon_F)}\right)^2\right)S^3\mu_B^2 H_{\text{ext}}^2\right),$$

which allows one to calculate the magnetic moment M

$$M = 2\mu_B^2\mathcal{N}(\varepsilon_F)SH_{\text{ext}}\left(1 + \frac{1}{6}\left(\frac{\mathcal{N}(\varepsilon_F)''}{\mathcal{N}(\varepsilon_F)} - 3\left(\frac{\mathcal{N}(\varepsilon_F)'}{\mathcal{N}(\varepsilon_F)}\right)^2\right)S^3\mu_B^2 H_{\text{ext}}^2\right). \tag{8.34}$$

In powers of the external field H_{ext} one finds that the next order term is proportional to H_{ext}^3. Although c is usually rather small, this effect is enhanced by the factor S^3 which is large for weak ferromagnets. The sign of c determines the direction of the deviation. Two cases are possible:

- $c < 0, \Rightarrow M$ shows a saturation behavior (Fig.8.7a);
- $c > 0$, $\frac{dM}{dH}$ diverges if H_{ext} becomes large enough. This causes a first order phase transition $\Rightarrow$ metamagnetism (Fig.8.7b).

8.4.6 Metamagnetism

One of the most prominent representatives of metamagnetism [91, 92] is YCo$_2$ [93, 94] which shows a first order phase transition to a magnetic state at a field of about 70T [95]. Metamagnetism is also present in similar alloys of the same structure type (cubic Laves phase [96]) like ScCo$_2$ and the pseudo binaries

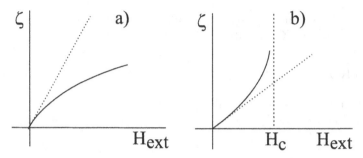

Fig. 8.7. Depending on the sign of c, the magnetic moment can show a saturation behavior (**a**) or a divergence (**b**)

Y(Co-Al)$_2$ [100], Sc(Co-Al)$_2$ [101], Lu(Co-Al)$_2$ [102]. In the paramagnetic state the Fermi energy is found near a minimum of the density of states, so that the case $c > 0$ is realized. For $\zeta = 0$ the Stoner criterion is not fulfilled since the free energy rises with increasing ζ. Only for larger values of ζ does the free energy drop again and a metastable minimum is found.

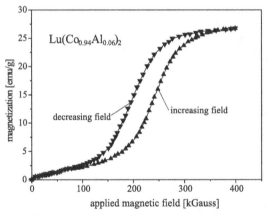

Fig. 8.8. Magnetization curves showing the metamagnetic transition (hysteresis) in Lu(Co$_{1-x}$Al$_x$)$_2$ for $x = 0.06$ at 4.2K

Figure 8.8 shows this behavior for the pseudo-binary cubic Laves phase Lu(Co$_{1-x}$Al$_x$)$_2$ with the characteristic non-linearity of the magnetization curve above the critical field H_{c2} [102]. The hysteresis found found for the magnetization is typical for a first order phase transition. The critical field can be calculated from the condition that the first and second derivative of $E(\zeta)$ with respect to ζ must vanish

$$H_{c2} = \frac{n}{3\mu_B \mathcal{N}(\varepsilon_F)} \frac{1}{\sqrt{cS^3}} \quad . \tag{8.35}$$

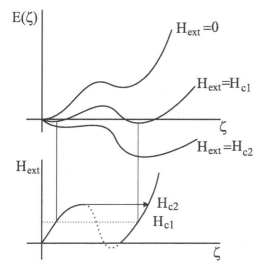

Fig. 8.9. Mechanism of a metamagnetic phase transition induced by an external field H_{ext}

Figure 8.9 explains how a metamagnetic transition happens. The zero field energy $E(\zeta)$ shows an anomaly for finite ζ. Note that in the limit $\zeta = 0$, the energy has an upwards curvature which means that the Stoner criterion is not fulfilled. For small moments the systems behaves like an ordinary non-magnetic one. Applying an external field tilts the energy curve around the origin because the external field enters the energy as $-H_{ext}\zeta$. For a field H_{c1} both possible minima have the same energy. At the field H_{c2} one minimum degenerates into a state where both the first and second derivative are zero. Consequently the susceptibility at this point diverges and the system undergoes a first order phase transition (discontinuous phase transition) to a magnetically ordered state for finite ζ.

One finds that the existence of a metamagnetic phase transition is very often coupled to a maximum in the temperature dependence of the susceptibility. The reason is again the similarity between the coefficients a in $\chi(T)$ (3.17) and c of $\chi(H)$ (8.24).

8.5 Susceptibility of Paramagnetic Alloys

As was shown in the previous section the Pauli susceptibility is enhanced by the exchange interaction. Susceptibility and magnetic moment in a thermodynamical sense are extensive variables (extensive variables scale with the volume) (e.g. magnetic field or temperature do not!). If one wants to determine the susceptibility per formula unit this can be done by taking the density of states per formula unit or of the unit cell

$$\chi_c = \sum_{i=1}^{n_c} \chi_i \quad , \tag{8.36}$$

where the index c denotes "cell".

In a more general approach to the problem defining a Stoner enhancement factor for an alloy, Jarlborg and Freeman [97] devised a method for calculating these values directly from the electronic band structure. They apply their formalism to the strongly enhanced Pauli paramagnet TiB$_2$ [98] and to the very weak itinerant magnet ZrZn$_2$ [99].

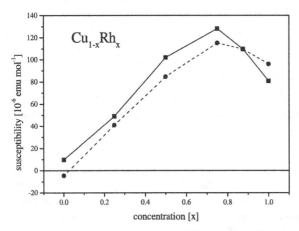

Fig. 8.10. Paramagnetic susceptibility of the alloy system Cu–Rh, calculation (full line) in comparison with experiment [104] (dashed line)

In the example given here, the susceptibility of the system Cu$_{1-x}$Rh$_x$ [103] (Fig. 8.10) is derived from the density of states (Fig. 8.11). Both pure Co and Rh have fcc crystal structure. The alloy system, however, has a miscibility gap between $0.2 \leq x \leq 0.9$ which leads to a phase segregation. By rapidly quenching from the melt one can stabilize an amorphous phase which has a number of nearest-neighbors close to 12 as in the fcc system. For $x = 0.75$ one finds a pronounced maximum in the susceptibility which must have its reason in the electronic structure.

If one calculates the densities of states (DOS) for an ordered fcc alloy one can explain this effect (Fig. 8.11). For $x = 1.0$ the DOS for fcc Rh is shown. It has the typical appearance for a $4d$-metal in the fcc structure with a not completely filled d-band. For Cu ($x = 0.0$) the $3d$-band is completely filled and the band width of the $3d$-band is already reduced with respect to fcc Ni for instance. The Fermi energy is in the region of the $4s$-band where the DOS is very small. This causes the Pauli susceptibility to be very small as well, so that it is exceeded by the diamagnetic contribution of the completely filled shells (Langevin diamagnetism) which shows up in experiment as a slightly negative

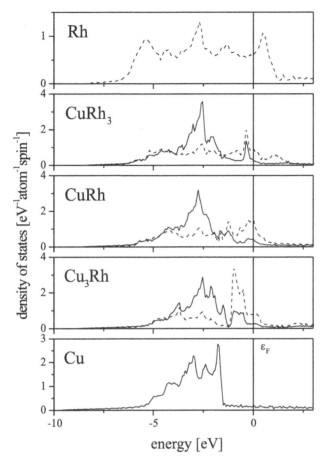

Fig. 8.11. Densities of states of the alloy system Cu–Rh. Cu-DOS (full line), Rh-DOS (dashed line)

value of χ for $x = 0$ (Fig. 8.10). On the Rh rich side the Fermi energy is found in a region of relatively high DOS leading to a large exchange enhanced susceptibility. For $CuRh_3$ the Fermi energy is just below a pronounced peak in the DOS. When, with further increase of the Rh concentration, the Fermi energy passes this peak, the susceptibility shows a peak as well which explains the experimental finding as being due to the peculiar electronic structure. The comparison with experiment shows a satisfactorily agreement.

9. Band Gap Theory of Strong Ferromagnetism

The Slater–Pauling (SP) curve (Fig. 5.1) shows two distinct branches with slopes ± 1. In the region of the positive slope one finds the weakly ferromagnetic systems, in the region of the negative slope the strongly ferromagnetic ones. How can one explain that behavior?

The total number of valence electrons Z and the magnetic moment M is given by two simple equations

$$Z = n^+ + n^- \quad , \quad M = n^+ - n^- \quad . \tag{9.1}$$

From these relations one can eliminate either n^+ or n^- and obtain

$$M = 2n^+ - Z \quad , \tag{9.2}$$

$$M = Z - 2n^- \quad . \tag{9.3}$$

Equation (9.2) accounts for the descending branch of the SP-curve: If the number of spin-up electrons n^+ is constant, which is the case for strong ferromagnetic systems, one finds a linear decrease of the magnetic moment with increasing number of valence electrons. Equation (9.2) thus describes the right, descending branch of the SP-curve. In the second case (9.3) one assumes that n^- is constant (which in fact means that n^- is zero), so that the magnetic moment rises with increasing Z. Equation (9.3) describes the left, ascending branch of the SP-curve. There remains the question of why either n^+ or n^- should remain constant. Basically Hund's first rule explains this feature. At first all spin-up electrons are added therefore n^- is constant because it is zero. Once one has added 5 electrons with spin-up (for a d-band) the remaining electrons can only be added with spin-down; now n^+ is constant because the spin-up band is already full. Ideally, for the first 5 electrons the magnetic moment rises linearly up to $5\mu_B$ and for the second 5 electrons it drops again linearly and becomes zero when the band is full. However, for itinerant systems one does not find $5\mu_B$ since Hund's rule is not strictly valid as discussed earlier.

The strongly ferromagnetic systems are those where the spin-up band is fully occupied. Here one notes that the Fermi energy is never found just at the upper edge of the spin up d-band, but is always separated by a gap $\triangle\varepsilon_s$. This gap is called the Stoner gap. If one assumes the validity of a rigid band model, the exchange interaction produces a transfer of electrons from spin-down to spin-up. This process comes to an end when the total energy

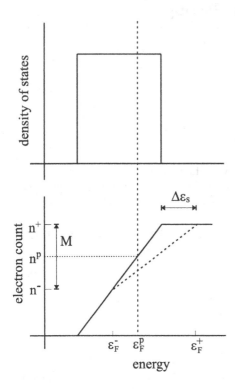

Fig. 9.1. Spin splitting for a strong ferromagnet with a rectangular band

is minimized. To study this process in detail one considers a d-band which is more than half filled. For simplicity a rectangular band is drawn in top panel of Fig. 9.1 (which by the way is close to the band shape of a closed packed transition metal).

In the bottom panel of Fig. 9.1 the integrated density of states, i.e. the number of electrons, is plotted. The creation of a magnetic moment is caused by the transfer of spin-down electrons to the spin-up band. Since with each spin-down electron leaving its original (spin-down) band exactly one spin-up electron is created, the resulting change in the occupation numbers is symmetrical about the paramagnetic occupation number n^p. As the bandwidth of the d-band is finite, the resulting energy splitting of a spin-up and spin-down Fermi energy is not symmetrical about the paramagnetic value.

From the minimum condition for the total energy one derives the following equilibrium condition for the spin splitting (compare to (8.11))

$$\triangle\varepsilon = \varepsilon^+ - \varepsilon^- = I_\mathrm{s} M \tag{9.4}$$

$$\Rightarrow \frac{M}{\varepsilon^+ - \varepsilon^-} = I_\mathrm{s}^{-1} = \frac{\mathrm{d}M}{\mathrm{d}\varepsilon} = H_\mathrm{M}^{-1} \quad .$$

The spin-splitting (band-splitting) comes to an end once the respective magnetic field is equal to the molecular field. In the case of the rectangular DOS this condition is given by the slope of the line connecting the points (ε^+, n^+) and (ε^-, n^-). Since this slope is also given by the averaged DOS between ε^+ and ε^- one can formulate the following mechanism for the formation of a magnetic moment: If the density of states is larger than I_{s}^{-1} the Stoner criterion is fulfilled and the splitting starts spontaneously. This process comes to an end when the equilibrium condition above is fulfilled. As long as both Fermi energies are in a region of a high density of states the slope is constant (for the rectangular band) and is given by $\frac{M}{\varepsilon^+ - \varepsilon^-}$. However, if the spin-up Fermi energy passes above the spin-up band the slope becomes smaller until finally the equilibrium condition is met.

The second case which will be investigated is one where the DOS is not of rectangular shape but shows a gap or at least a valley (even in the latter case the word gap will be used, although strictly speaking a gap is an energy interval where the DOS is zero). Assuming that the DOS shows a pronounced minimum which is of parabolic shape (top panel of Fig. 9.2) the same derivation as above is applied. One finds that the existence of a minimum causes the slope $\frac{M}{\varepsilon^+ - \varepsilon^-}$ to vary rapidly. Such a gap is thus able to stabilize the position of the spin-up Fermi energy. If, as in Fig. 9.2 (bottom panel) the paramagnetic Fermi energy lies below the minimum of the DOS, the spin-up Fermi energy will end up at a position slightly above the minimum. This effect is also known as pinning. Further electrons can only enter the spin-down band and thus reduce the net magnetic moment. This case describes a quasi strongly ferromagnetic behavior on the descending branch of the Slater–Pauling curve.

If the paramagnetic Fermi energy lies above the minimum, the spin down energy becomes pinned and one describes a weakly ferromagnetic system. Further electrons can only enter the spin-up band and the magnetic moment increases. This is an example for the ascending branch of the Slater–Pauling curve.

This latter case is found for bcc Fe. The minimum in the DOS pins the spin-down Fermi energy and causes the weak itinerant behavior of iron and the "too small" magnetic moment of $2.2\mu_{\mathrm{B}}$. Band structure calculation of the fcc Fe phase show that, for enlarged volume, fcc Fe would have a magnetic moment of $\cong 2.6\mu_{\mathrm{B}}$. The reason is the different DOS for a fcc structure which does not show the pronounced gaps as for the bcc case. In fcc Fe the Fermi energy does not become pinned in a gap, so that fcc Fe is a strong ferromagnet like its right hand neighbors in the periodic table, hcp Co and fcc Ni. Such calculations of seemingly hypothetical structures are not in the least academic. Although fcc Fe does not exist as a bulk material, it can be produced as an epitaxially grown thin film [105, 106] or as a precipitate in a copper matrix [107]. In both cases the measured magnetic moment agrees fairly well with the result from the band structure calculation.

At this point one has to redefine the terms "weak" and "strong" ferromagnetism. Following our investigations one finds strong ferromagnetism, whenever the Fermi energy is in a region of low DOS. The lower the DOS is, the stronger the ferromagnetism becomes. The reason is that the DOS determines the susceptibility (8.30) which is the response function of the system. If the susceptibility is small also the response of the system to an external perturbation such as field, pressure etc. is small because the magnetic moment cannot follow the perturbation. The occurrence of the Fermi energy in a gap or above the spin-up band, leads to a similar strongly ferromagnetic behavior [108].

At this point it becomes clear why bcc Fe behaves so strangely. According to the older definition bcc Fe would be a weak ferromagnet because the spin-up band is not completely filled. From experiment one would expect bcc Fe to be more on the stronger side, because the magnetic moment is large so that the susceptibility is not too high and the pressure dependence of the Curie temperature and the magnetic moment are very small. Only the position of the Fermi energy in a gap can explain this situation so that bcc Fe is near to the transition from a weakly to strongly ferromagnetic system and is thus found on the top of the Slater–Pauling curve, where both branches intersect.

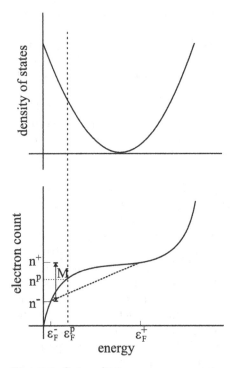

Fig. 9.2. Spin splitting over a gap state

9.1 Magnetism of Alloys

In the following section it is demonstrated that the magnetic moment of an alloy between a magnetic host lattice (mostly Fe, Co, or Ni) and a dissolved constituent, the solute, (mostly non-magnetic transition metals or metalloids) can easily be calculated if strongly ferromagnetic behavior is present. One considers an alloy of the composition $A_{1-x}B_x$, where A is the host and B the solute. The average moment per atom is given by

$$\mu_{\mathrm{av}} = \mu_A (1-x) + \mu_B (x) = n^+ - n^- \quad . \tag{9.5}$$

The average valency, i.e. the number of electrons outside the last completely filled shell, is the sum of spin-up plus spin-down electrons

$$Z_{\mathrm{av}} = Z_A (1-x) + Z_B (x) = n^+ + n^- \quad . \tag{9.6}$$

Eliminating n^+ or n^- one obtains as before

$$\mu_{\mathrm{av}} = 2\left(n_{\mathrm{sp}}^+ + n_{\mathrm{d}}^+\right) - Z_{\mathrm{av}} \quad , \quad \mu_{\mathrm{av}} = Z_{\mathrm{av}} - 2\left(n_{\mathrm{sp}}^- + n_{\mathrm{d}}^-\right) \quad . \tag{9.7}$$

The number of valence electrons are now split into a contribution due to the s and p states n_{sp} and one due to the d states n_{d}. Assuming that either n_{sp}^+ and n_{d}^+ or n_{sp}^- and n_{d}^- is constant under alloying one finds

$$\mu_{\mathrm{av}} = \mu_A^0 - x\left(Z_B - Z_A\right) \quad , \quad \mu_A^0 = 2\left(n_{\mathrm{sp}}^+ + n_{\mathrm{d}}^+\right) - Z_A \quad , \tag{9.8}$$

$$\mu_{\mathrm{av}} = \mu_A^0 + x\left(Z_B - Z_A\right) \quad , \quad \mu_A^0 = Z_A - 2\left(n_{\mathrm{sp}}^- + n_{\mathrm{d}}^-\right) \quad , \tag{9.9}$$

where μ_A^0 is the host moment. Again one describes the change of slope distinguishing the two branches of the Slater-Pauling curve.

As an interlude a brief historical review is given here. The relations above were formulated for the first time by N.F. Mott [27] in 1935. For conventional strong ferromagnets he assumed that the spin-up band is filled so that $2n^+ = 10$. Mott also assumed that n_{sp} is constant but he had severe difficulties justifying this assumption. This simple model led to useful results for Ni-Cu, Ni-Zn, Ni-Co, Co rich Fe-Co, and Ni rich Fe-Ni alloys, all of which are found on the descending branch of the SP curve. The second form (9.9) explains the weakly ferromagnetic systems like Fe-Cr and Fe rich Fe-Co where one assumes that the rigid band model is valid and Co and Cr (in these alloys) form similar band structures to bcc Fe (an effect which is called "common band behavior"). If the Fermi energy is pinned in a gap, n_{d}^- stays constant and one describes the ascending branch of the SP-curve (see Fig. 5.1). This model explains, for instance Fe-Co, Fe-Cr, and Fe-V.

Despite this success, the rigid band model ran into trouble, because with the publication of the first band structure calculations it became clear that the bands of the solute were far from rigid. This led to criticism of the rigid band model formulated as: "success based on canceling mistakes".

The first modification which accounted for the new band structure results was formulated by J. Friedel [109]. He found out that in alloys between constituents at the beginning or the end of the transition metal series, the strongly repulsive potential of the host causes a solute state close to the Fermi energy which contains room for 10 electrons (impurity state). If the Fermi energy is again pinned in the gap between the host d -band and the impurity peak, Friedel modified the equations accordingly

$$\mu_{av} = \mu_A^0 - x\left(10 + Z_B - Z_A\right) \quad . \tag{9.10}$$

This generalization now explains most of the side branches appearing in the SP-curve such as Co-Cr, Co-V, Ni-Cr, and Ni-V.

Although the models satisfactorily explain the magnetic moments of transition metal alloys it is not clear how one has to alter the theory to describe alloys between a magnetic transition metal and a non-magnetic metalloid such as B, Al, or Si. If one replaces a strongly ferromagnetic transition metal by a metalloid one effectively reduces the number of spin-up d-electrons by 5 while at the same time the number of sp-electrons should stay constant. This is questionable insofar as from the position in the periodic table one would expect that those metalloids which occur on the right hand side of the periodic table should have fully occupied s-bands and partially occupied p-states. It is thus surprising that experimental data of Ni-Al, Ni-Ge, Ni-Si actually confirm the fixed number of sp-electrons.

The Ni alloys became famous for a different reason. By chance Ni has exactly 10 valence electrons. This reduces the Friedel formula (9.10) to[1]

$$\mu_{av} = \mu_A^0 - xZ_B \quad . \tag{9.11}$$

How can one understand the constancy of the n_{sp} electron number and why is it possible that metalloid atoms which have a large number of s- and p-electrons leave this n_{sp} number unchanged?

When one adds metalloid atoms one has to consider two effects which are able to create sp-electrons in the host:

i) The attractive potential of the metalloid shifts unoccupied sp-states below the Fermi energy where they become occupied. This effect would lead to an increase in the n_{sp} number and is called "new state filling" and is a covalent bonding which occurs in metals and alloys.

ii) The alternative process is "polarization" of the occupied states of the host. States in the host are composed of linear combinations of states of nearest-neighbors. These are already occupied, so that this mechanism does not increase the number of sp-states.

From band structure calculations one finds that the latter case occurs much more often than the first one, thus the number of n_{sp} remains unchanged. The reason is found in the so called hybridization gap which is caused by the interaction between the sp- and the d-states of the host. Even if the potential of the solute is attractive it would cost a lot of energy to move

---

[1] There exist a number of publications where this formula is used for other systems as well (as a counterexample to the rigid band picture) although it is only valid for Ni alloys with B atoms from the beginning of the transition metal series.

states across that gap. Thus this mechanism is suppressed in favor of the polarization. The hybridization gap is thus the reason that the sp-potential is shielded by the polarization. This effect is also called Fano–*anti-resonance*. This terminology is chosen in analogy of the p- and d-resonance (Wigner delay time [110, 119]) which leads to peaks in the density of states. Figure 9.3 shows this anti-resonance for the intermetallic compound CoAl.

Having found an argument why n_{sp} should stay constant one can go back to our original problem. One defines a *magnetic valence* [111] $Z_{\mathrm{m,av}}$ by

$$Z_{\mathrm{m,av}} = \sum_i x_i Z_\mathrm{m}^i \quad , \text{with} \quad Z_\mathrm{m}^i = 2n_{\mathrm{d}i}^+ - Z^i \quad . \tag{9.12}$$

In the case of a binary alloy this gives

$$Z_{\mathrm{m,av}} = (1-x)\left(2n_{\mathrm{dA}}^+ - Z_\mathrm{A}\right) + x\left(2n_{\mathrm{dB}}^+ - Z_\mathrm{B}\right) \quad . \tag{9.13}$$

The average magnetic moment is then

$$\mu_{\mathrm{av}} = 2n_{\mathrm{sp}}^+ + Z_{\mathrm{m,av}} \quad . \tag{9.14}$$

The magnetic valence as defined by (9.12) of Fe, Co, Ni, Cu, Zn are $2, 1, 0, -1, -2$. The respective values for the metalloids B, Si, P are $-3, -4, -5$.

As an example the magnetic moments of the transition metal borides MnB, FeB, CoB are calculated and compared to experiment; $n_{\mathrm{sp}}^+ = 0.5$:

MnB: $Z_{\mathrm{m,av}} = 0.5 \times 3 + 0.5 \times (-3) = 0$ $\mu_{\mathrm{av}} = 1 + 0 = 1\mu_\mathrm{B}$
FeB: $Z_{\mathrm{m,av}} = 0.5 \times 2 + 0.5 \times (-3) = -0.5$ $\mu_{\mathrm{av}} = 1 - 0.5 = 0.5\mu_\mathrm{B}$
CoB: $Z_{\mathrm{m,av}} = 0.5 \times 1 + 0.5 \times (-3) = -1$ $\mu_{\mathrm{av}} = 1 - 1 = 0\mu_\mathrm{B}$

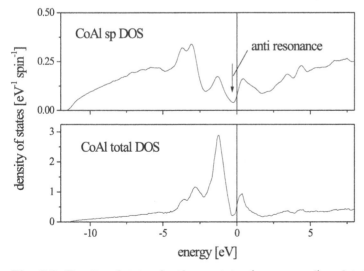

Fig. 9.3. Density of states for the sp-states (upper panel) and total DOS (lower panel). In both panels the DOS is given per formula unit CoAl

Since the metalloid atom (boron) does not carry a magnetic moment, the "average" moment μ_{av} must reside entirely at the transition metal. This yields $\mu_{Mn} = 2\mu_B$, $\mu_{Fe} = 1\mu_B$, and $\mu_{Co} = 0\mu_B$. If one compares these values with experiment one finds a fair agreement: MnB: $1.83\mu_B$/Mn-atom, FeB: $1.12\mu_B$/Fe-atom, CoB is non-magnetic [112].

This simple method of determining the magnetic moment of an alloy between a magnetic and a non-magnetic constituent can of course never replace the experiment or a full bodied electronic structure calculation. However, it provides a tool to acquire a certain feeling about what size of a moment can be expected. This feeling proves to be rather helpful when it comes to the interpretation or evaluation of experimental or theoretical results.

10. Magnetism and the Crystal Structure – Covalent Magnetism

Since the carriers of magnetism in the transition metals are the d-electrons it is obvious that the description of the metallic properties via the free electron model has its limitations. The d-electrons are much more tightly bound than the s-electrons and they form directional bonds. A picture like this is typical for a covalently bound solid. However, metals are not completely covalently bound (like diamond) since they show metallic conductivity. A characterization which probably describes the covalency in metals best was given by V. Heine: "metals are systems with unsaturated covalent bonds" [113]. This characterization not only account for the directional (covalent) bonds which are responsible for the crystal structure but also for the typical metallic properties which distinguish metals from insulaters. The electrons which partly occupy the orbitals formed by the covalent interaction have a relatively high "resonance" energy which makes it easy for them to hop from atom to atom as required for a true metal. This "chemical" picture can even be applied to explain where the particular shape of the DOS in metals comes from. As an example the schematic DOS of the d-electrons in a bcc metal (like nonmagnetic iron) will be derived. Figure 10.1 shows the local environment of an atom in the bcc structure. The central atom is surrounded by a nearest-neighbor (n.n.) shell of 8 atoms (black circles) at a distance of $a_0\sqrt{3}/2$ and a next-nearest-neighbor (n.n.n.) shell of 6 atoms (grey circles) at a distance of a_0.

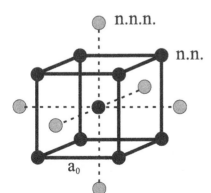

Fig. 10.1. Nearest-neighbor (n.n.) and next-nearest-neighbor (n.n.n) shells in a bcc structure. a_0 is the lattice constant

To describe the interactions for this cluster one has to distinguish between stronger n.n. interactions along the shorter n.n.-distance and weaker n.n.n. interactions. The left panel of Fig. 10.2 shows the interaction of two free atoms using a molecular-orbital (MO) diagram. When the two atoms are brought together, their atomic wavefunctions start to overlap and form hybrid orbitals. Within a tight binding model (Chap. 4), these hybrid orbitals can be described as linear combinations of Wannier functions. These hybrid orbitals split into a bonding and an anti-bonding (marked by $*$) orbital. The width of the splitting depends on the energy difference of the respective atomic states and for most on the spatial separation of the two atoms. The splitting becomes large if the energy separation is small and the atoms are close together, and vice versa. In the present case of a bcc environment, there exist 8 neighbors at a closer distance of $a_0\sqrt{3}/2$. Their atomic wave functions overlap strongly and form the wide split bottom and top hybrid orbitals. The n.n.n. shell is located at a distance of a_0. Consequently the overlap between the atomic wavefunctions is smaller and the resulting energy splitting of the hybrid orbitals as well. This interaction leads to the two orbitals in the center of the MO-diagram. Each atom now provides a number of electrons which fill the newly formed hybrid orbitals.

The MO-diagram describes the interactions still on a molecular level. Assuming that a periodic solid is formed of a periodic array of such molecules, it is easy to understand that the energetic sequence of the hybrid orbitals will also be present in the solid. However, the molecular energy levels will form bands and thus will be broadened according to the strength of the interactions. The right panel of Fig. 10.2 shows the transformation of the

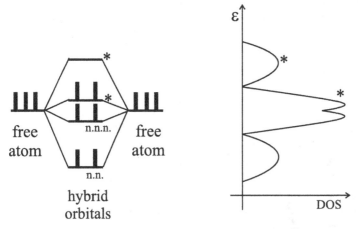

Fig. 10.2. Left panel: MO-diagram for the interaction of 2 atoms in a bcc environment; Right panel: transformation of the energy levels of the hybrid orbitals into a density of states (DOS)

MO-diagram into the respective density of states of a periodic solid with bcc structure. The DOS consists of two major features:

- The strong n.n. interaction leads to a broad peak (bonding states) at low energy and a similar one at the top of the DOS (anti-bonding states).
- In the center of the DOS one finds a double peak which stems from the weaker n.n.n. interaction. Also this peak consists of a bonding and an anti-bonding component at its bottom and top energy, respectively. Due to the broadening of the band states, these bonding and anti-bonding states largely overlap and form the pronounced double peak.

This central peak is a peculiar feature of the bcc structure. Since it consists of a mixture of bonding and anti-bonding states its contribution to the chemical bond is rather weak so that it is often also referred to as "transition metal non-bonding peak". The DOS which has been derived here should directly be compared to the result of a full band structure calculation as shown in Fig. 8.5 for bcc Fe. Such calculation yields almost exactly the shape which has been derived on the basis of the MO-diagram. The fact that the shape of the DOS of the d-electrons of a solid can be described along these line justifies the earlier statement about the covalent interactions present in a metallic solid. The common picture of a metallic state which consists of a positively charged lattice of atomic cores and the freely moving valence electrons obviously breaks down once the much more localized d-electrons enter the game. The discussion given here is of course not restricted to non-magnetic solids. It is easy to generalize this picture to magnetically split states where atomic orbitals of like spin will interact to form spin dependent hybrid orbitals.

10.1 Crystal Structure of Mn, Fe, Co, and Ni

Comparing iso-electronic elements one finds that these elements not only have comparable chemical properties but also group together in their crystal structure. This tendency to form groups of like crystal structure is not only restricted to the transition metals but holds, with very few exceptions, for the whole periodic table [114]. Figure 10.3 shows an excerpt from the periodic table concentrating on the s-d transition metal series. All three series show the same structural trend as a function of the number of valence electrons: hcp→bcc→hcp→fcc. This behavior can be traced back to the electronic structure and the interaction of the s- and d-band in these elements [117]. The magnetic elements Mn, Fe, and Co however, deviate from this general trend. Mn which would be expected to be hcp, forms a highly complicated structure type (α-Mn structure) with 58 atoms per unit cell. Also its magnetic structure is extremely complicated exhibiting anti-ferromagnetism together with non-collinear spin ordering. The most stable form of Fe is the bcc structure although Fe would actually fall into the hcp group. At $T = 0$K Co is hcp and not fcc as expected. One should note that Co shows a crystallographic phase transition towards the fcc structure already below the Curie temperature. This suggests that both the hcp and the fcc modification in Co must have comparable energies. This is not surprising, since the two close packed structures (hcp and fcc) are closely related to each other (both structures have 12 nearest neighbors and differ only in the stacking sequence along the [111] direction). The reason why Mn, Fe, and Co fall out of the general trend must thus be caused by their magnetic order. The number of valence electrons $n^\uparrow + n^\downarrow$ is the sum of the s- and the d-electrons. Assuming a spin-up and a spin-down band $n^\uparrow$ and $n^\downarrow$ are the same for the non-magnetic elements. For the magnetic elements magnetism causes a band splitting and thus a different occupation for the spin-up and the spin-down band. If one assumes that, for simplicity, the spin-up d-band is full with 5 electrons (like in a strong ferromagnet) the remaining electrons occupy the spin-down band which makes 2, 3, and 4 spin-down electrons for Mn, Fe, and Co, respectively (see the grey shaded box in Fig. 10.3).

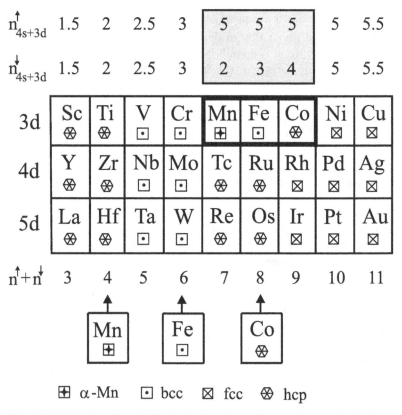

Fig. 10.3. Explanation of the deviations from the usual crystal structure for Mn, Fe, and Co

From the theory of the chemical bond it is known that filled electron shells do not contribute to the bonding. This means that for Mn, Fe, and Co the full spin-up band can be neglected and the electronic and structural properties are governed by the partially filled spin-down band. Since the crystal structure is determined by the number of valence electrons in the partially filled band, Mn, Fe, and Co behave as if they would have 2, 3, and 4 valence electrons which are responsible for the chemical bond and thus for the crystal structure. Fe thus falls in the Cr, Mo, W group and thus adopts the bcc structure. Co falls in the Ru, Os group and becomes hcp (note that the fcc region is however very close) [115]. Mn, as mentioned above, is rather a case in itself. The fact that for the magnetic elements the spin-up electrons do not contribute to the bonding also leads to a significant reduction of the bulk modulus. Concomitantly one also observes an increase in volume, an effect which is known as volume magnetostriction.

10.2 Covalent Magnetism

Another direction of approach to the problem of the magnetism of alloys is covalent magnetism[116]. As an example, the system Fe-Co is chosen. The two metals are neighbors in the periodic table which makes them very similar concerning their electronic structure. They also have almost the same electro-negativity so that any substantial charge transfer between them can be ruled out. Pure Fe and pure Co crystallize bcc and hcp, respectively. $Fe_{1-x}Co_x$ for $x \leq 0.8$ also crystallizes in a body centered structure forming an alloy where the Fe and Co atoms randomly occupy the lattice sites. On alloying they will form a common band as long as the energies of the respective Fe- and the Co-states are comparable. If one were to just assume a rigid band behavior one would simply fill this bcc band with the necessary number of electrons. This would mean however that one allows a charge transfer of half an electron from cobalt to iron. This charge transfer, as mentioned above, is completely unphysical, not only due to the comparable electronegativities but also due to the metallic character of the bond. In addition because of electrostatics, a charge transfer is energetically always very "expensive". However if one performs a band structure calculation one finds that only for the fully occupied spin-up band the common band picture is valid (where one has the same number of electrons for both spin-up Co and spin-up Fe namely 5). This is not the case for the spin-down bands and there it is where covalent magnetism occurs. Fig. 10.4 shows how this interaction can be understood from a simple molecular orbital (MO) scheme. One starts from the non-spinpolarized states of Fe and Co. Due to the additional electron the Co states will be lower in energy than the Fe ones. If one now allows for a band splitting, Fe will show a larger splitting that Co. Since only states with like spin can interact, Fe and Co spin up, which are about at the same energy will form a common band. For the spin down states, an energy difference

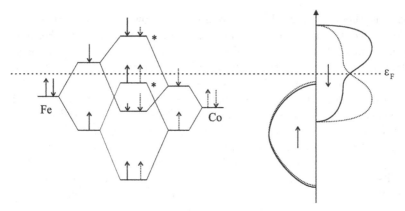

Fig. 10.4. MO diagram of the interaction in FeCo and its translation into the density of states

occurs. When these states form a molecular (hybrid) orbital, the occupation with electrons will be inversly proportional to the energy difference between the atomic level and the MO-level. This means that for the spin down bonding MO more Co that Fe electrons will be transferred and vice versa for the antibonding state (this process is sketched by the variable length of the spins arrows). If this picture is now translated into a density of states (right panel of Fig. 10.4), it leads to an unperturbed DOS for spin up (the Co-DOS, dotted, is slightly shifted to make it visible) and a distortion for the spin down DOS according to the MO mechanism described before.

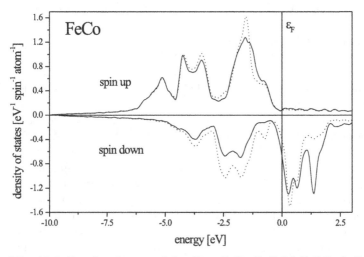

Fig. 10.5. Density of states of the alloys FeCo. Fe-DOS (full line), Co-DOS (dotted line)

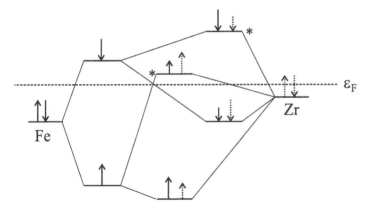

Fig. 10.6. Schematic view of the interaction in ZrFe$_2$ leading to the ferrimagnetic coupling of the Zr moment

Fig. 10.5 shows an actual DOS from a calculation of the electronic band structure. One identifies exactly the mechanism described for Fig. 10.4 with a common band for the spin up DOS and the redistribution of spectral weight in the spin down band. Since this redistribution enhances the number of Co states below ε_F, Co can accommodate its additional electron without the need of a charge transfer.

Another example of covalent magnetism is found in the cubic Laves phase compounds e.g. ZrFe$_2$ [118] (Fig. 10.6). Here an intermetallic compound between a magnetic and a non- magnetic partner is formed. One again starts from the non-magnetic states, where Zr, because of its small number of valence electrons will be higher in energy than Fe. The peculiar energetic position of the Fe and Zr states causes a magnetic moment at Zr which is antiparallel to the moment on Fe although pure Zr does not show a spontaneous magnetic moment. The occupation of the resulting MO's follows the arguments given in the description above. The result can also be used to determine the mutual magnetic polarization, by just adding up the length of the spin-up and spin-down arrows for the occupied states (below ε_F). One finds that the moment of Fe is reduced with respect to pure Fe and that at Zr a magnetic moment is formed which is antiparallel to the Fe moment. The mechanism demonstrated here for ZrFe$_2$ has been generalized as "covalent polarization" [161]. One finds that for an alloy between a magnetic and a non-magnetic partner, the polarization on the non-magnetic site is anti-parallel for an atom from the beginning of a series in the periodic table (small number of valence electrons). Changing the non-magnetic alloy partner through the series results in a progressive lowering of the anti-parallel polarization until in the second half of the series a parallel polarization occurs.

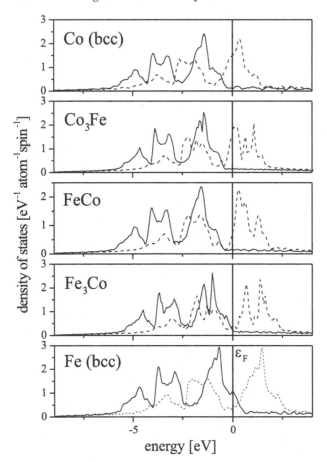

Fig. 10.7. Densities of states for the ordered phases of the alloy system Fe-Co. The DOS shown is the total DOS per unit cell and spin divided by the number of atoms per unit cell (normalized to one atom); spin-up DOS: full line, spin-down DOS: dashed line

As a final example again the alloy system Fe-Co is discussed. Up to about 80% Co, Fe-Co forms a body centered alloy and for higher Co concentrations a transition to the hcp structure is observed. In the example given here, the system Fe-Co will be simulated by assuming ordered structures of the bcc type. For Fe_3Co and Co_3Fe the Heusler structure was assumed. In this structure which is of the general type ABX_2 there exist always either two types of Fe atoms (Fe_3Co) of two types of Co atoms (Co_3Fe) which differ by their nearest-neighbor coordination. In Fig. 10.8 these local coordination shells are shown explicitly. Figure 10.7 shows the DOS for the systems bcc Fe, Fe_3Co (Heusler structure), FeCo (CsCl-structure), $FeCo_3$ (Heusler structure), and hypothetical bcc Co. (The hypothetical bcc Co structure was also calculated to simulate a complete miscibility). Already in Fe_3Co the Fermi energy

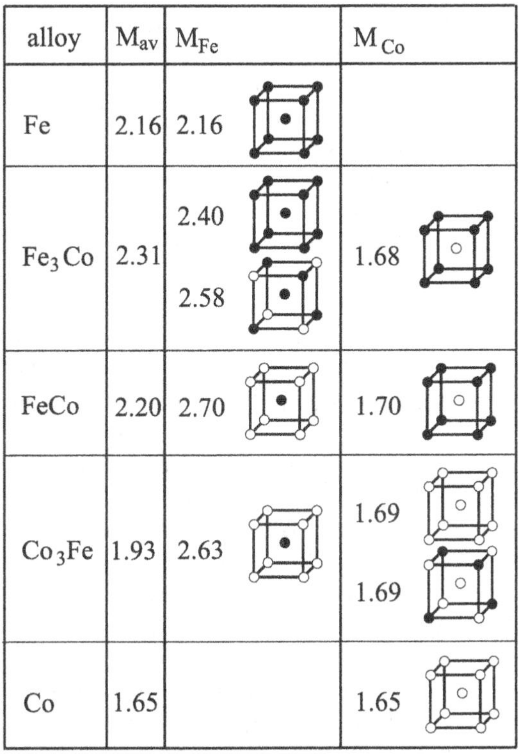

alloy	M_{av}	M_{Fe}		M_{Co}	
Fe	2.16	2.16			
Fe$_3$Co	2.31	2.40 2.58	 	1.68	
FeCo	2.20	2.70		1.70	
Co$_3$Fe	1.93	2.63		1.69 1.69	
Co	1.65			1.65	

Fig. 10.8. Magnetic moments for Fe and Co in the alloys system Fe-Co (after [119]). It is found that the magnetic moment for Fe depends strongly on the type of neighboring atoms, whereas the moment of Co does not

is pushed above the spin-up band but remains pinned in a minimum of the spin-down band. The system is close to a transition to a strong ferromagnet and the magnetic moment per unit cell is a maximum.

In FeCo the splitting is again increased but for Fe the maximum moment is reached for $2.6 - 2.7 \mu_B$ being at the border line of a strongly ferromagnetic system. One finds that the magnetic moment of Co remains unchanged. In [119] a second modification of the FeCo alloy has been studied as well. This is the Zintl-phase (NaTl-structure) where each atom is surrounded by 4 neighbors of its own and 4 of the other kind. Fe which lies between the weakly and strongly ferromagnetic regime changes its moment according to the coordination having the smaller moment when being surrounded by 8 atoms of its own kind thus being "closer" to bcc Fe.

In FeCo$_3$ the moment of both alloy partners is saturated and one finds a strong ferromagnet. The magnetic moment of Co is as large as in pure (hcp) Co. This constancy of the magnetic moment of Co demonstrates once more that for a strongly ferromagnetic system the magnetic moment does not depend on the crystal structure and on the chemical environment.

It is also interesting to ask how the magnetization density (this is the difference between the spin-up and the spin-down electron density) is distributed within a magnetic solid. Such an analysis can be drawn from neutron diffraction experiments or, nowadays much more easily, from calculations of the electronic structure. Figures 10.9 and 10.10 shows the result of a full potential LMTO (Linear Muffin Tin Orbital) calculation for the ordered phase of FeCo in the CsCl structure. It can immediately be seen that the magnetization density is far from being uniformly distributed over the whole crystal as one would expect for an itinerant electron magnet. However, this is not really surprising, since the carriers of the magnetic moments are the d-electrons which are certainly not delocalized themselves. This result again demonstrates that the assumption of free electrons in a metal is over idealizing the actual situation when it comes to the transition metals. Since the magnetization density is produced by the comparably narrow $3d$-bands is appears to be fairly localized around the atomic positions. But there is one more fact to be noticed: From the contour-plot (Fig.10.10) one notices that the spin density, and also the charge density, is strongly aspherical. This asphericity is particularly pronounced for Co. This feature is again a direct consequence of the covalent interactions present in the FeCo alloy. However, at this point of the book the reader should no longer be surprised that the magnetic transition metals are not very good examples for true itinerant electron ferromagnets. After all, one hardly knows better ones.

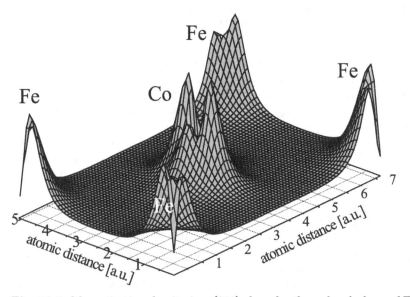

Fig. 10.9. Magnetization density in a [011] plane for the ordered phase of FeCo

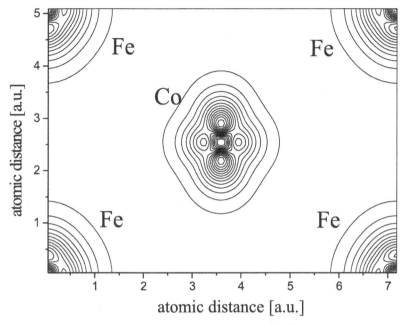

Fig. 10.10. Magnetization density contours in an [011] plane for the ordered phase of FeCo. The asphericity of the spin density around the Co atom is clearly visible

10.3 Covalent Polarization

When a magnetic atom is inserted into a non-magnetic host lattice, the atoms of the host become polarized by this magnetic impurity and show a magnetic moment. In the limiting case of the free electron gas, this polarization is described by the RKKY interaction (see Chap. 11). In a more realistic description of a solid, the electronic states of the magnetic impurity atom will interact with the neighboring atoms of the host. Since this interaction can again be described on the basis of a molecular orbital picture, one can derive a model which is based on the covalent interaction of the impurity with the host electronic states. As long as the concentration of the magnetic impurity atoms is low, these atoms will form fairly localized states, so that these impurities behave like almost free atoms. The host lattice remains almost unperturbed by the presence of these impurities, apart from the magnetic polarization described herein. In Fig. 10.11 the assumed energetic scheme is shown. The electronic structure of the non-magnetic host is described by a density of states, which for simplicity is assumed to be of rectangular shape $\mathcal{N}(\varepsilon) = $ const. The localized impurity atom is assumed to have two atomic like energy levels $E^\uparrow$ and $E^\downarrow$ for spin-up and spin-down electrons, respectively. This magnetic splitting occurs symmetrically around a non-magnetic state E_0. The spin-split energy levels $E^\uparrow$ and $E^\downarrow$ interact with the respective states of the host band h_{22} from the bottom of the host band up to the Fermi

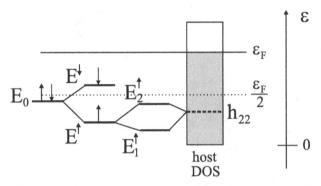

Fig. 10.11. Energy level scheme to determine covalent polarization. The host states are drawn as a rectangular shaped density of states (grey shaded rectangle), the impurity state E_0 is split magnetically into $E^{\uparrow}$ and $E^{\downarrow}$ which interact with the host state. For simplicity only the covalent interaction between $E^{\uparrow}$ and h_{22} is shown

energy ε_F. The strength of this interaction will be described by a single parameter h_{12} which in fact is given by the overlap between the respective wave functions of the impurity atom and the host atoms. In the picture of covalency, this interaction leads to a splitting of the energy levels $E^{\uparrow}$ and $E^{\downarrow}$ into the eigenvalues $E_1^{\uparrow}$, $E_2^{\uparrow}$ and $E_1^{\downarrow}$, $E_2^{\downarrow}$ which are given by

$$E_1^{\uparrow\downarrow} = \frac{E^{\uparrow\downarrow} + h_{22}}{2} - \frac{1}{2}\sqrt{\left(E^{\uparrow\downarrow} - h_{22}\right)^2 + 4h_{12}^2} \quad,$$

$$E_2^{\uparrow\downarrow} = \frac{E^{\uparrow\downarrow} + h_{22}}{2} + \frac{1}{2}\sqrt{\left(E^{\uparrow\downarrow} - h_{22}\right)^2 + 4h_{12}^2} \quad. \tag{10.1}$$

If $\Psi_{\text{imp}}\left(E^{\uparrow\downarrow}\right)$ is the respective wave function of the impurity atom for the energy $E^{\uparrow\downarrow}$ and $\Psi_{\text{host}}\left(h_{22}\right)$ is the wave function of the host, the total wave function Φ can be written as a linear combination of $\Psi_{\text{imp}}\left(E^{\uparrow\downarrow}\right)$ and $\Psi_{\text{host}}\left(h_{22}\right)$ which reads

$$\Phi = C_A\Psi_{\text{imp}}\left(E^{\uparrow\downarrow}\right) + C_B\Psi_{\text{host}}\left(h_{22}\right) \quad. \tag{10.2}$$

The total energy E_{tot} of the interacting system is calculated from $E_{\text{tot}} = \langle\Phi|H|\Phi\rangle$ where H is the Hamiltonian describing the interaction giving

$$E_{\text{tot}} = C_A^*C_A \underbrace{\langle\Psi_{\text{imp}}|H|\Psi_{\text{imp}}\rangle}_{E^{\uparrow\downarrow}} + C_B^*C_B \underbrace{\langle\Psi_{\text{host}}|H|\Psi_{\text{host}}\rangle}_{h_{22}}$$

$$+ C_A^*C_B \underbrace{\langle\Psi_{\text{imp}}|H|\Psi_{\text{host}}\rangle}_{h_{12}} + C_B^*C_A \underbrace{\langle\Psi_{\text{host}}|H|\Psi_{\text{imp}}\rangle}_{h_{21}} \quad, \tag{10.3}$$

where its was assumed that both Ψ_{imp} and Ψ_{host} are normalized to 1. C_A and C_B are the linear combination coefficient which themselves obey the normalization condition $C_A^2 + C_B^2 = 1$. Minimizing the total energy with respect to the two linear combination coefficients leads to a set of homogeneous linear equations

$$\frac{\partial E_{\text{tot}}}{\partial C_A^*} = C_A E^{\uparrow\downarrow} + C_B h_{12} = 0$$

$$\frac{\partial E_{\text{tot}}}{\partial C_B^*} = C_A h_{21} + C_B h_{22} = 0 \quad . \tag{10.4}$$

The secular equation of this problem is given by the determinant of the coefficients of (10.4). Due to the hermicity of the problem $h_{12} = h_{21}$ so that the secular equation reads

$$\left(E^{\uparrow\downarrow} - \lambda \right) \left(h_{22} - \lambda \right) - h_{12}^2 = 0 \tag{10.5}$$

The two solutions for λ are the energies given by (10.1) Form this result C_A and C_B become

$$C_A^2 \left(E_1^{\uparrow\downarrow} \right) = \frac{h_{12}^2}{h_{12}^2 + (E^{\uparrow\downarrow} - E_1^{\uparrow\downarrow})^2} \quad , \tag{10.6}$$

$$C_A^2 \left(E_2^{\uparrow\downarrow} \right) = \frac{h_{12}^2}{h_{12}^2 + (E^{\uparrow\downarrow} - E_2^{\uparrow\downarrow})^2} \quad , \tag{10.7}$$

$$C_B^2 \left(E_1^{\uparrow\downarrow} \right) = \frac{(E^{\uparrow\downarrow} - E_1^{\uparrow\downarrow})^2}{h_{12}^2 + (E^{\uparrow\downarrow} - E_1^{\uparrow\downarrow})^2} \quad , \tag{10.8}$$

$$C_B^2 \left(E_2^{\uparrow\downarrow} \right) = \frac{(E^{\uparrow\downarrow} - E_2^{\uparrow\downarrow})^2}{h_{12}^2 + (E^{\uparrow\downarrow} - E_2^{\uparrow\downarrow})^2} \quad . \tag{10.9}$$

The interaction will modify the density of states of the host in a different way for the spin-up and spin-down states. The magnetic moment which results from this interaction is given by the difference of the spin-up and spin-down host electrons $N_{\text{host}}^{\uparrow}$ and $N_{\text{host}}^{\downarrow}$

$$N_{\text{host}}^{\uparrow} = \int_0^{\varepsilon_F} \frac{\mathcal{N}(h_{22})}{2} \left[C_B^2 \left(E_1^{\uparrow} \right) \frac{dE_1 \left(E^{\uparrow} \right)}{dh_{22}} + C_B^2 \left(E_2^{\uparrow} \right) \frac{dE_2 \left(E^{\uparrow} \right)}{dh_{22}} \right] dh_{22} \quad , \tag{10.10}$$

$$N_{\text{host}}^{\downarrow} = \int_0^{\varepsilon_F} \frac{\mathcal{N}(h_{22})}{2} \left[C_B^2 \left(E_1^{\downarrow} \right) \frac{dE_1 \left(E^{\downarrow} \right)}{dh_{22}} + C_B^2 \left(E_2^{\downarrow} \right) \frac{dE_2 \left(E^{\downarrow} \right)}{dh_{22}} \right] dh_{22} \quad . \tag{10.11}$$

Putting everything together one obtains for the magnetic moment of the host lattice $M_{\text{host}} = N_{\text{host}}^{\uparrow} - N_{\text{host}}^{\downarrow}$

$$M_{\mathrm{host}} = \int\limits_0^{\varepsilon_{\mathrm{F}}} \frac{\mathcal{N}(h_{22})}{2}$$

$$\times \frac{\left(E^\uparrow - E^\downarrow\right)\left(E^\uparrow + E^\downarrow - 2h_{12}\right) h_{12}^2 \, dh_{22}}{\left(E^{\uparrow 2} + 4h_{12}^2 - 2E^\uparrow h_{22} + h_{22}^2\right)\left(E^{\downarrow 2} + 4h_{12}^2 - 2E^\downarrow h_{22} + h_{22}^2\right)}.$$

$$(10.12)$$

Carrying out the integration and considering our assumption of a rectangular density of states $\mathcal{N}(\varepsilon) = \mathcal{N}(\varepsilon_{\mathrm{F}}) = $ const. yields

$$M_{\mathrm{host}} = \frac{h_{12}}{2}\mathcal{N}(\varepsilon_{\mathrm{F}}) \left[\arctan\left(\frac{E^\downarrow}{2h_{12}}\right) - \arctan\left(\frac{E^\uparrow}{2h_{12}}\right) \right.$$
$$\left. - \arctan\left(\frac{E^\downarrow - \varepsilon_{\mathrm{F}}}{2h_{12}}\right) + \arctan\left(\frac{E^\uparrow - \varepsilon_{\mathrm{F}}}{2h_{12}}\right)\right] \quad .$$

Expanding the $\arctan(x) = x - \frac{x^3}{3} + \ldots$ and identifying the spin splitting of the impurity atoms as ΔE given by $\Delta E = E^\uparrow - E^\downarrow$ and $E^\uparrow + E^\downarrow = 2E_0$ yields

$$M_{\mathrm{host}} = -\frac{\mathcal{N}(\varepsilon_{\mathrm{F}})\varepsilon_{\mathrm{F}}}{8h_{12}^2}\Delta E\left(E_0 - \frac{\varepsilon_{\mathrm{F}}}{2}\right) \quad . \tag{10.13}$$

Equation (10.13) describes the polarization of a host lattice by a magnetic impurity and is called "covalent polarization" [161]. It is physically intuitive that the magnetic moment scales with the spin-splitting ΔE and thus with the impurity moment. If one expresses the spin splitting ΔE in terms of the impurity moment M_{imp} and the respective Stoner factor $I_{\mathrm{S}}^{\mathrm{imp}}$ as $\Delta E = I_{\mathrm{S}}^{\mathrm{imp}} M_{\mathrm{imp}}$ one finds a linear relation between the band filling expressed in terms of E_0 and the polarization ratio $M_{\mathrm{host}}/M_{\mathrm{imp}}$.

$$\frac{M_{\mathrm{host}}}{M_{\mathrm{imp}}} = -\frac{\mathcal{N}(\varepsilon_{\mathrm{F}})\varepsilon_{\mathrm{F}}}{8h_{12}^2} I_{\mathrm{S}}^{\mathrm{imp}}\left(E_0 - \frac{\varepsilon_{\mathrm{F}}}{2}\right) \quad . \tag{10.14}$$

What is surprising is the fact that the sign of the polarization depends on the energetic position of the impurity states with respect to the Fermi energy. For impurities with a less than half filled shell, E_0 will be larger than $\varepsilon_{\mathrm{F}}/2$ so that the resulting host polarization will be antiparallel to the impurity moment. For an impurity atom with a half filled shell E_0 will become about equal to $\varepsilon_{\mathrm{F}}/2$ so that M_{host} will be close to zero. Consequently for impurity atoms with a more than half filled shell the host polarization is parallel to the impurity moment. This behavior is shown in Fig. 10.12 where results from an ab initio band structure calculation of the "giant moment" system Pd-TM (TM=transition metal) are plotted as a function of the type of impurity atom (=position of E_0) [161]. For a Cr impurity with a less than half filled shell the resulting polarization is negative. For Mn with its half filled d shell, although its magnetic moment is close to $4\mu_{\mathrm{B}}$, the polarization is close to

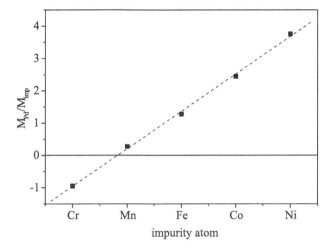

Fig. 10.12. Polarization ratio between the host moment M_{Pd} per cell and the impurity moment M_{Pd} as a function of the impurity atom. The data are taken from calculations of a $Pd_{31}TM$ supercell, where TM = Cr, Mn, Fe, Co, Ni. The dotted line is a linear least square fit to the data

zero. Ni with its almost completely filled shell causes the large polarization although its magnetic moment is only $0.84\mu_B$.

Calculations of the magnetic structure of such impurity systems confirm the simple linear dependence of the host polarization as a function of the band filling [120, 121]. This result can also be compared to earlier investigations which arrive at similar conclusions [122]–[125].

11. Magnetic Impurities in an Electron Gas

J. Friedel [109] noted that an impurity put in a gas of free electrons (jellium) causes an oscillatory perturbation of the electron density around it, which shields the impurity. The charge density oscillations are called Friedel oscillations. The physical origin for this perturbation is the scattering of the free electrons at the impurity potential. The magnetic analogue is the Rudermann, Kittel, [126] Kasuya [127], and Yoshida [128] (RKKY) interaction which describes the interaction between a localized magnetic impurity and the surrounding electron gas. The RKKY interaction caused by the superposition of the charge density oscillations of the spin-up and the spin-down electrons giving rise to a spin density oscillation.

11.1 Impurity Potential in the Jellium

In the jellium model of a solid it is assumed that the positive charge of the cores is continuously smeared out to form a constant positive background. The electron density is thus also constant and the respective eigenfunctions are plane waves $\exp(i\mathbf{kr})$. If one inserts an impurity potential $V_0(\mathbf{r})$ it will become shielded by the electron gas. The resulting shielded *effective* potential $V(\mathbf{r})$ will lead to a change in the electron density $\Delta n(\mathbf{r})$. In lowest order perturbation theory one can write this density change as [129]

$$\Delta n(\mathbf{r}) = \int \chi_0(\mathbf{r}, \mathbf{r}') V(\mathbf{r}') \, d\mathbf{r}' \quad . \tag{11.1}$$

One now calculates the susceptibility $\chi_0(\mathbf{r}, \mathbf{r}')$. It should be noted that this is not yet a magnetic susceptibility. The term susceptibility is just used in its more general meaning as a response function. To calculate $\chi_0(\mathbf{r}, \mathbf{r}')$ one first introduces the Green function $G(\mathbf{r}, \mathbf{r}'; E)$ to the Hamiltonian H. In its spectral representation $G(\mathbf{r}, \mathbf{r}'; E)$ can be expressed using the eigenvalues E_α and the eigenfunctions ψ_α to H

$$G(\mathbf{r}, \mathbf{r}'; E) = \sum_\alpha \frac{\psi_\alpha(\mathbf{r}) \psi_\alpha(\mathbf{r}')}{E + i\epsilon - E_\alpha} \quad . \tag{11.2}$$

The charge density is then given by

$$n\left(r\right) = 2 \sum_{\alpha, E_\alpha < E_F} |\psi_\alpha\left(r\right)|^2 = -\frac{2}{\pi} \int^{E_F} \mathrm{Im}\left\{G\left(r, r; E\right)\right\} dE \quad . \tag{11.3}$$

For a small perturbation, Dyson's equation $G = G_0 + G_0 V G$ is approximated by $G \simeq G_0 + G_0 V G_0$ which allows to calculate the Green function of the perturbed system G from the Green function of the unperturbed system G_0

$$\Delta n\left(r\right) = -\frac{2}{\pi} \int^{E_F} dE\, \mathrm{Im} \int G_0\left(r, r'; E\right) V\left(r\right) G_0\left(r', r; E\right) dr' \quad , \tag{11.4}$$

so that the susceptibility defined in (11.1) reads

$$\chi\left(r, r'\right) = -\frac{2}{\pi} \int^{E_F} dE\, \mathrm{Im}\left\{G_0\left(r, r'; E\right) G_0\left(r', r; E\right)\right\} \quad . \tag{11.5}$$

In the case of free electrons $\left(H_0 = -\frac{\hbar^2}{2m}\partial_r^2\right)$ the unperturbed Green function is a spherical wave of the form

$$G_0\left(r, r'; E\right) = G_0\left(r - r'; E\right) = -\frac{2m}{\hbar^2 4\pi} \frac{\exp\left(ik\,|r - r'|\right)}{|r - r'|} \quad , \tag{11.6}$$

$$\text{with} \quad k = \frac{k}{|k|}\sqrt{\frac{2m}{\hbar^2}E} > 0 \quad .$$

The resulting susceptibility is easy to calculate and its radial part reads

$$\chi\left(R\right) = \frac{m}{\hbar^2\left(2\pi\right)^3} \frac{1}{R^4} \left(2k_F R \cos\left(2k_F R\right) - \sin\left(2k_F R\right)\right)$$

$$\simeq \frac{m}{\hbar^2\left(2\pi\right)^3} \frac{2k_F \cos\left(2k_F R\right)}{R^3} \quad for \quad 2k_F R \gg 1 \quad . \tag{11.7}$$

The susceptibility $\chi\left(R\right)$ and therefore also the density oscillate with a period π/k_F (for distances far enough from the perturbation). The amplitude of these oscillations decays with $\frac{1}{R^3}$. These oscillations are a direct consequence of the oscillatory character of the free Green function (i.e. a spherical wave). Apart from the formal derivation one can also easily give a classical explanation of this effect. If a plane wave representing the free electron hits a point target (the scattering potential) a spherical wave is emitted from that target. This spherical wave is exactly the free Green function. The classical optics analogue would simply be Huygen's principle.

11.2 Strong Perturbations in the Jellium

For strong perturbations linear response theory is no longer valid. In this case the potential V must be replaced by the energy dependent T-Matrix

$$T\left(E\right) \equiv V \frac{1}{1 - G_0\left(E\right)V} \quad . \tag{11.8}$$

This expression is a consequence from the Dyson equation

$$G = G_0 + G_0VG$$
$$= G_0 + G_0VG_0 + G_0VG_0VG_0 + \cdots \quad ,$$

which formally can be solved by introducing the T-Matrix to be

$$G = G_0 + G_0TG_0 \quad . \tag{11.9}$$

The result for the density oscillations reads

$$\Delta n\left(r\right) \simeq \frac{2m}{\hbar^2\left(2\pi\right)^3} \frac{k_F\left|t\right|}{r^3} \cos\left(2k_F r + \delta\right) \quad , \tag{11.10}$$

where $\left|t\right|$ is the trace of the T-matrix at E_F.

One finds that a strong perturbation leads to the same type of oscillations but there occurs a phase shift which depends on the strength of the potential.

11.3 Layer and Line Defects

With the advancement of the necessary experimental techniques, the physics of magnetic superstructures became a central object of investigation. In particular structures which consist of repeated magnetic and non-magnetic layers show new features like *giant magneto resistance* (GMR) which are of great technological interest. If one wants to describe the magnetic coupling between such layers one has to consider perturbations which are of one- and two-dimensional shape. The derivation is analogous so that only the result is given.

For a layer in the x, y plane the susceptibility along the z direction is given by

$$\chi^{(2)}\left(z\right) \simeq -\frac{m}{\hbar^2\left(2\pi\right)^2} \frac{\sin\left(2k_F z\right)}{z^2} \quad . \tag{11.11}$$

For a line defect the perturbation potential $V\left(r\right) = V\left(\rho\right)$; $\rho = \sqrt{x^2 + y^2}$ independent of z. The result is

$$\chi^{(1)}(\rho) \simeq \frac{m(2k_{\mathrm{F}})^3}{\hbar^2(2\pi)^3}\left(\frac{\cos(2k_{\mathrm{F}}\rho)}{\rho^{\frac{5}{2}}}a + \frac{\sin(2k_{\mathrm{F}}\rho)}{\rho^{\frac{5}{2}}}b\right) \quad , \qquad (11.12)$$

$$\text{with}\quad a = \int_0^\infty du\,\frac{\cos u}{\sqrt{u}} \quad , \quad b = \int_0^\infty du\,\frac{\sin u}{\sqrt{u}} \quad ,$$

so that the oscillation decays as $\frac{1}{\rho^{5/2}}$.

11.4 Magnetic Impurities and Oscillations of the Magnetization

In many non-magnetic host lattices $3d$-impurities possess a magnetic moment. Very famous are the $3d$-impurities in a palladium host which have large moments (almost at the Hund's rule limit) and polarize the surrounding Pd atoms to form a large magnetization cloud. This oscillatory polarization can be understood from the charge oscillations discussed above. Following density functional theory spin-up and spin-down electrons "feel" a different potential of the form

$$V_\pm(\boldsymbol{r}) \simeq v_c(\boldsymbol{r}) \pm v_x(\boldsymbol{r})\,m(\boldsymbol{r}) \quad , \qquad (11.13)$$

which means that spin-up and spin-down electrons are scattered differently. The main difference is the different phase shift which gives rise to oscillations of the spin-up electron density shifted relative to the oscillations of the

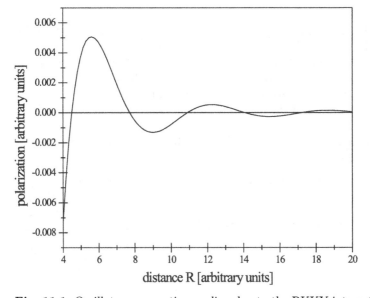

Fig. 11.1. Oscillatory magnetic coupling due to the RKKY interaction

spin-down density. The superposition of these two charge densities yields an oscillatory magnetization which decays according to the dimensionality of impurity considered. Fig. 11.1 depicts this behavior following (11.7). This RKKY-interaction is the reason that atoms at a given distance from the impurity either feel a positive or negative polarization and consequently have magnetic moments of respective orientation. The RKKY-interaction is also employed to explain the behavior of canonical spin-glasses. A spin-glass is a system (e.g. small amounts of Fe dissolved in a Au host) where localized magnetic moments exists which are oriented at random. Since these orientations appear to be "frozen" below a certain temperature, in analogy to a undercooled amorphous melt the name spin-glass was coined. It is assumed that any magnetic impurity (which themselves are distributed at random) produces an RKKY polarization. These polarizations interfere with each other and thus produce a completely random polarization pattern in which the impurity moments are locked. It is easy to understand that such systems have highly degenerate ground states which only differ in the orientation of the impurity moments. Consequently one often finds extremely slow relaxation processes which can happen on a time scale of months or even years.

12. Itinerant Electrons at $T > 0$: A Historical Survey

The introduction of finite temperature to the Stoner model caused problems from the very beginning. The temperature dependences predicted are too weak and, even worse, the analytic behavior of the temperature dependent variables do not agree with experiment. Finally there were only a few systems which could be described within this model (e.g. $ZrZn_2$, Ni_3Al). These systems must have an extremely high susceptibility and a very small magnetic moment, consequently their Curie temperatures are very small as well. For all other systems the Stoner model has proved to be hardly applicable and at best provides a correction to the collective excitations which dominate the finite temperature properties. However, the model became very famous at the time and that fact makes it worthwhile going into the details.

As in the chapter about the temperature dependence of the paramagnetic susceptibility one assumes that the temperature dependence of the Fermi distribution function is the crucial quantity. Using (3.6) one writes the sum of states

$$Z = \frac{1}{2}n \left(\frac{k_B T}{\varepsilon_F} \right)^{\frac{3}{2}} \left(F_{3/2}\left(\eta + \beta + \beta'\right) + F_{3/2}\left(\eta - \beta - \beta'\right) \right) \quad , \qquad (12.1)$$

where

$$\beta' = \frac{\mu_B H_{ext}}{k_B T} \quad , \quad \eta = \frac{\mu}{k_B T} \quad , \quad \beta = \frac{k_B \Theta \zeta}{k_B T} \quad . \qquad (12.2)$$

β' describes the influence of an external field and β the influence of the molecular field. In the case when the external field is zero β' vanishes. The magnetization per atom is then given by

$$M = \frac{3}{4}n\mu_B \left(\frac{k_B T}{\varepsilon_F} \right)^{\frac{3}{2}} \left(F_{1/2}\left(\eta + \beta\right) - F_{1/2}\left(\eta - \beta\right) \right) \quad ,$$

$$n = \frac{3}{4}n \left(\frac{k_B T}{\varepsilon_F} \right)^{\frac{3}{2}} \left(F_{1/2}\left(\eta + \beta\right) + F_{1/2}\left(\eta - \beta\right) \right) \quad , \qquad (12.3)$$

and the relative magnetization ζ becomes

$$\zeta = \frac{F_{1/2}\left(\eta + \beta\right) - F_{1/2}\left(\eta - \beta\right)}{F_{1/2}\left(\eta + \beta\right) + F_{1/2}\left(\eta - \beta\right)} = \frac{n^+ - n^-}{n^+ + n^-} \quad . \qquad (12.4)$$

Equation (12.4) is a rather complicated relation for ζ which is given as a function of T and via the characteristic temperature Θ describing the molecular field. However, given the density of states, the number of spin-up and spin-down electrons can be written

$$n^\pm = \frac{n}{2}(1 \pm \zeta)$$

$$= \int_0^\infty \mathcal{N}(\varepsilon)\left(\exp\left(\frac{\varepsilon}{k_B T} - \eta^\pm\right) + 1\right)^{-1} d\varepsilon \quad, \tag{12.5}$$

with

$$\eta^\pm = k_B T \eta \pm k_B \Theta \zeta \pm \mu_B H_{ext} \quad, \tag{12.6}$$

which are the famous Stoner equations.

These are now used to determine the paramagnetic susceptibility. By means of the Stoner equations the magnetization M is given by

$$M = n \mu_B \zeta$$

$$= \mu_B \left[\int_0^\infty \mathcal{N}(\varepsilon)\left(\exp\left(\frac{\varepsilon}{k_B T} - \eta + \beta + \beta'\right) + 1\right)^{-1} d\varepsilon \right.$$

$$\left. - \int_0^\infty \mathcal{N}(\varepsilon)\left(\exp\left(\frac{\varepsilon}{k_B T} - \eta - \beta - \beta'\right) + 1\right)^{-1} d\varepsilon \right]$$

$$= \mu_B \left[G(\eta + \beta + \beta') - G(\eta - \beta - \beta') \right]$$

$$= 2\mu_B (\beta + \beta') \frac{dG}{d\eta} = 2\mu_B \frac{\mu_B H_{ext} + k_B \Theta \zeta}{k_B T} \frac{dG}{d\eta} \quad,$$

where $G(\eta)$ is given by

$$G(\eta) = \int_0^\infty \mathcal{N}(\varepsilon) f(\varepsilon) d\varepsilon$$

$$= \int_0^\infty \mathcal{N}(\varepsilon)\left(\exp\left(\frac{\varepsilon}{k_B T} - \eta\right) + 1\right)^{-1} d\varepsilon \quad,$$

$$\Rightarrow \frac{dG}{d\eta} = -k_B T \int_0^\infty \mathcal{N}(\varepsilon) \frac{df(\varepsilon)}{d\varepsilon} d\varepsilon = k_B T \int_0^\infty \mathcal{N}(\varepsilon) \left|\frac{df(\varepsilon)}{d\varepsilon}\right| d\varepsilon \quad.$$

One thus obtains for the magnetic moment

$$M = 2\mu_B \left(\mu_B H_{ext} + k_B \Theta \frac{M}{n\mu_B}\right) \int_0^\infty \mathcal{N}(\varepsilon) \left|\frac{df(\varepsilon)}{d\varepsilon}\right| d\varepsilon \quad, \tag{12.7}$$

and for the susceptibility

$$\chi = \frac{M}{H_{\text{ext}}}$$

$$= \frac{2\mu_B^2 \int\limits_0^\infty \mathcal{N}(\varepsilon) \left|\frac{\mathrm{d}f(\varepsilon)}{\mathrm{d}\varepsilon}\right| \mathrm{d}\varepsilon}{1 - 2\frac{k_B\Theta}{n} \int\limits_0^\infty \mathcal{N}(\varepsilon) \left|\frac{\mathrm{d}f(\varepsilon)}{\mathrm{d}\varepsilon}\right| \mathrm{d}\varepsilon} . \tag{12.8}$$

For $T = 0$, (12.8) can be traced back to the result given in (8.31).

One can now study the inverse susceptibility $1/\chi$ for various values of the molecular field $(k_B\Theta)/(n\mu_B)$

$$\frac{1}{\chi} = \left(2\mu_B^2 \int\limits_0^\infty \mathcal{N}(\varepsilon) \left|\frac{\mathrm{d}f(\varepsilon)}{\mathrm{d}\varepsilon}\right| \mathrm{d}\varepsilon\right)^{-1} - \frac{k_B\Theta}{n\mu_B^2} . \tag{12.9}$$

The temperature dependence of $1/\chi$ is shown in Fig. 12.1. All three curves are parallel. The curvature, which is roughly quadratic does not give a Curie–Weiss law and there are only few systems which show this dependence (e.g. Ni-Al, Ni-Ga, Ni-Pt). Curve "0" is the function without a molecular field being the temperature dependence of the non-interacting susceptibility. Curve "1" describes systems which are on the verge of magnetic order, which means that $\chi = \infty$ at $T = 0$. From this curve one can again derive the Stoner criterion

$$0 = \frac{1}{2\mu_B^2 \mathcal{N}(\varepsilon_F)} - \frac{k_B\Theta}{n\mu_B^2} ,$$

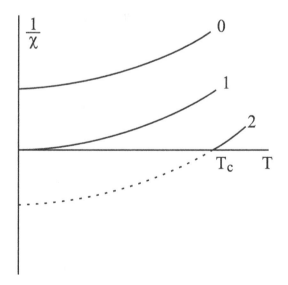

Fig. 12.1. Temperature dependence of the inverse susceptibility in the Stoner model

$$\Rightarrow 2\mathcal{N}\left(\varepsilon_{\mathrm{F}}\right)\frac{k_{\mathrm{B}}\Theta}{n} = 1 \quad .$$

For curve "2" there is a critical temperature above which the susceptibility becomes positive. This critical temperature is the Curie temperature in the Stoner model. From the condition that $1/\chi$ must be zero at T_{c} one derives

$$\frac{2k_{\mathrm{B}}\Theta}{n}\int\limits_{0}^{\infty}\mathcal{N}\left(\varepsilon\right)\left|\frac{\mathrm{d}f\left(\varepsilon\right)}{\mathrm{d}\varepsilon}\right|\mathrm{d}\varepsilon = 1 \quad \text{at } T_{\mathrm{c}} \quad . \tag{12.10}$$

Equation (12.10) is nothing other than a temperature dependent Stoner criterion. Due to the convolution with the function $\mathrm{d}f\left(\varepsilon\right)/\mathrm{d}\varepsilon$, the DOS at the Fermi energy drops with rising temperature until the lhs of (12.10) becomes equal to 1. This temperature is then called the Curie temperature. It should be noted here that this definition of the Curie temperature means that the magnetic moments at each atom vanishes individually. This assumption is basically different from the picture of statistical disorder or a breakdown of the long range correlation which appear in the Weiss and the Heisenberg model.

Earlier on the temperature dependence of the integral in (12.10) was calculated. Using (3.18) one obtains

$$2\mathcal{N}\left(\varepsilon_{\mathrm{F}}\right)\frac{k_{\mathrm{B}}\Theta}{n}\left(1 + aT_{\mathrm{c}}^2\right) \quad , \tag{12.11}$$

$$\text{with} \quad a = \frac{\pi^2}{6}k_{\mathrm{B}}^2\left(\frac{\mathcal{N}\left(\varepsilon_{\mathrm{F}}\right)''}{\mathcal{N}\left(\varepsilon_{\mathrm{F}}\right)} - \left(\frac{\mathcal{N}\left(\varepsilon_{\mathrm{F}}\right)'}{\mathcal{N}\left(\varepsilon_{\mathrm{F}}\right)}\right)^2\right) \quad ,$$

which is an equation to determine T_{c} given that $a < 0$. The constant a has already been identified as an effective Fermi degeneracy temperature $a = T_{\mathrm{F}}^{-2}$ (3.18). For $T_{\mathrm{c}} \ll T_{\mathrm{F}}$ one writes (12.11) as

$$k_{\mathrm{B}}\Theta = \frac{n}{2\mathcal{N}\left(\varepsilon_{\mathrm{F}}\right)}\left(1 + \frac{T_{\mathrm{c}}^2}{T_{\mathrm{F}}^2}\right) \quad . \tag{12.12}$$

Using (8.27) one can equate the respective terms to those in (12.12) and finds

$$\frac{1}{3}\left|c\right|\zeta^2 = \left|a\right|T_{\mathrm{c}}^2 \tag{12.13}$$

Equation (12.13) shows that in the Stoner model, the Curie temperature scales linearly with the magnetic moment. This linear relation is typical for Fermi liquid theories. In Chap. 18 where spin fluctuations are introduced it will be found that T_{c} becomes proportional to M^2.

From (8.31) one also derives the result

$$\frac{1}{3}\left|c\right|\zeta^2 = \left|a\right|T_{\mathrm{c}}^2 = 2\mu_{\mathrm{B}}^2\mathcal{N}\left(\varepsilon_{\mathrm{F}}\right)\chi^{-1} = 2\mathcal{N}\left(\varepsilon_{\mathrm{F}}\right)\frac{k_{\mathrm{B}}\Theta}{n} - 1 \quad . \tag{12.14}$$

If one inspects the relations in (12.14) more closely and for that purpose again introduces the Stoner factor I_{s}, the Curie temperature of an itinerant ferromagnet within the Stoner model is given by

$$T_c^2 = T_F^2 \left(I_s \mathcal{N} \left(\varepsilon_F \right) - 1 \right) \quad . \tag{12.15}$$

Equation (12.15) relates the Fermi degeneracy temperature T_F and the Curie temperature T_c . Since T_F is usually of the order of several thousand Kelvin one only obtains reasonable values for T_c if $\left(I_s \mathcal{N} \left(\varepsilon_F \right) - 1 \right)$ becomes very small. This means that Stoner theory will only be applicable for very weak itinerant systems but will fail for most of the everyday magnetic materials, which is actually the case.

12.1 Excitations at Low Temperatures

From the Stoner equations (12.5) and (12.6) one can calculate the thermal excitation of the magnetization ζ at low temperatures. Again the two cases of weak and strong ferromagnetism are discussed.

12.1.1 Strongly Ferromagnetic Systems

The situation around the Fermi energy of a strong ferromagnet (the spin-up band is completely filled) is depicted in Fig. 12.2.

Here $k_B T \eta_0^-$ is the chemical potential of the spin-down electrons, $\Delta E = 2k_B \Theta$ is the band splitting (for a strong ferromagnet $\zeta = 1$). $\Delta = \Delta E - k_B T \eta_0^-$ is the Stoner gap. At $T = 0$ one has

$$\Delta + k_B T \eta_0^- = 2k_B \Theta \quad , \quad \Delta + k_B T \eta_0^+ = 0 \quad , \tag{12.16}$$

and

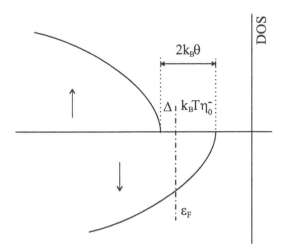

Fig. 12.2. The upper band edge of a strong ferromagnet

$$n = n^- = \int\limits_0^{k_B T \eta_0^-} \mathcal{N}(\varepsilon)\, d\varepsilon \quad , \quad n^+ = 0 \quad . \tag{12.17}$$

At low temperatures the softening of the Fermi distribution function being of the order $k_B T$ is much smaller than the Stoner gap. In this case one can approximate Fermi statistics by the classical Boltzmann distribution, which yields a simple mechanism [134] describing the thermal excitation of spin-up electrons in the unoccupied states of the spin-down band.

$$n^+ = \frac{n}{2}(1 - \zeta)$$

$$= \int\limits_0^\infty \mathcal{N}(\varepsilon) \exp\left(-\frac{\varepsilon}{k_B T}\right) \exp\left(-\frac{\Delta}{k_B T}\right) d\varepsilon$$

$$= F(T) \exp\left(-\frac{\Delta}{k_B T}\right) \quad . \tag{12.18}$$

Equation (12.18) describes a thermally induced "weakening" of the strong ferromagnet. Electrons are transferred from the majority to the minority band. The magnetization is thus

$$\zeta = 1 - \frac{2}{n} F(T) \exp\left(-\frac{\Delta}{k_B T}\right) \quad . \tag{12.19}$$

One can now use experiments to determine the various parameters. For fcc Ni it was shown [132] that $F(T)$ obeys a relation of the form

$$\frac{2}{n} F(T) = 2.6 \times 10^{-5} T^{\frac{3}{2}} \tag{12.20}$$

It is not surprising that the low temperature behavior of Ni has the same form as it was found for spinwave excitations. This behavior exists at low temperature for almost all magnetic systems, since the collective excitations which are responsible for it can already easily excited at low temperatures.

Taking also the Stoner gap into account one finds a highly accurate fit function for the magnetization of Ni

$$\zeta = 1 - 2.6 \times 10^{-5} T^{\frac{3}{2}} \exp\left(-\frac{440}{T}\right) \tag{12.21}$$

This gives a Stoner gap of about 40meV. The values given in the literature scatter greatly, e.g. neutron diffraction data give 60meV. The proof of the existence and the value of the Stoner gap was once a great challenge for all kinds of spectroscopic methods ranging from neutrons to photo electron spectroscopy.

12.1.2 Weakly Ferromagnetic Systems

The derivation for weakly ferromagnetic systems is somewhat more complicated because one has to consider the framework of the Sommerfeld expansions which were also used to determine the effect of metamagnetism (see Sect. 8.4.6). One starts at $T = 0$ where the magnetic moment ζ_0 is given by

$$\frac{n}{2}\left(1 \pm \zeta_0\right) = \int_0^\infty \mathcal{N}\left(\varepsilon\right) d\varepsilon \quad .$$

Using the results derived earlier one writes

$$1 - \frac{\zeta\left(T\right)}{\zeta_0} = \frac{3}{2c^*\zeta_0^2} a^* T^2 \quad , \tag{12.22}$$

where a^* and c^* are given by

$$a^* = \frac{\frac{1}{6}\pi^2 k_B^2}{2k_B\Theta\zeta_0^2} \left[\frac{\mathcal{N}\left(\varepsilon^+\right)'}{\mathcal{N}\left(\varepsilon^+\right)} - \frac{\mathcal{N}\left(\varepsilon^-\right)'}{\mathcal{N}\left(\varepsilon^-\right)}\right] \quad , \tag{12.23}$$

and

$$c^* = \frac{\frac{3}{2}}{2k_B\Theta\zeta_0^2} \left[2k_B\Theta - \frac{n}{2}\left(\frac{1}{\mathcal{N}\left(\varepsilon^+\right)} + \frac{1}{\mathcal{N}\left(\varepsilon^-\right)}\right)\right] \quad . \tag{12.24}$$

One finds that the magnetization shows a T^2 dependence, which is very weak. (Fe: $1 - \frac{\zeta}{\zeta_0} = 1.9 \times 10^{-6}$ at $4K$) Using (8.30) one can express c^* in terms of the interacting susceptibility

$$c^* = -2\frac{3}{2}\frac{1}{2k_B\Theta\zeta_0^2}\frac{n\mu_B^2}{\chi} \quad . \tag{12.25}$$

For $\zeta_0 \ll 1$ one replaces the susceptibility by means of (8.32)

$$\frac{n\mu_B^2}{\chi} = \frac{1}{3}\frac{n|c|\zeta_0^2}{\mathcal{N}\left(\varepsilon_F\right)} \quad , \tag{12.26}$$

$$\Rightarrow c^* = -|c|\frac{n}{2k_B\Theta\mathcal{N}\left(\varepsilon_F\right)} \simeq -|c| \quad .$$

Now the behavior of (12.22) is determined by the sign of a^*. Depending on the sign of a^* one has to distinguish two cases:

1. $a^* < 0$ and $c^* < 0$: The magnetization decreases with increasing temperature following a T^2 behavior

$$\left(\frac{\zeta}{\zeta_0}\right)^2 = 1 - \frac{T^2}{T_c^2} \quad . \tag{12.27}$$

2. $a^* > 0$ and $c^* < 0$: In this extremely unlikely case one should observe a magnetic moment which increases with rising temperature. This effect has once been seen in Y_2Ni_7 [130]. A few years later M. Shimizu [131] even provided a theoretical model. However, later investigations were not able to confirm the original finding. Today it is believed that an additional ferromagnetic phase contained in the sample caused these surprising results.

12.2 Stoner Theory for a Rectangular Band

To calculate the magnetization $\zeta = \zeta(T, \Theta, H)$ one must solve the Stoner equations (12.5) and (12.6). In his original work Stoner solved these equations for the parabolic (free electron like) band. During the 1970s calculations of more complicated forms appeared in the literature. It was found that for realistic values of Θ the shape of the magnetization curves depends strongly on the details of the density of states. To get a feeling for such a calculation one now considers a very simple but useful band shape, the rectangular band where $\mathcal{N}(\varepsilon) = \text{const.}$

For a band of the shape depicted above the density of states is given by

$$\mathcal{N}(\varepsilon) = \frac{n}{2\varepsilon_F} \quad . \tag{12.28}$$

Putting this expression into the Stoner equations yields ($H = 0$)

$$\frac{n}{2}(1 \pm \zeta) = \frac{n}{2\varepsilon_F} \int_0^\infty \left[\exp\left(\frac{\varepsilon}{k_B T} - \eta \pm \frac{k_B \Theta \zeta}{k_B T} \right) + 1 \right]^{-1} d\varepsilon \quad ,$$

with the abbreviations

$$x = \frac{\varepsilon}{k_B T} \quad , \quad \rho = \frac{k_B \Theta \zeta}{k_B T} \quad ,$$

one obtains

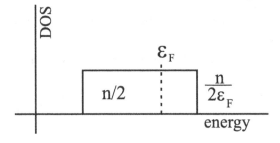

Fig. 12.3. Rectangular model density of states

$$\varepsilon_F \left(1 \pm \zeta\right) = k_B T \int\limits_0^\infty \left[\exp\left(x - \eta \pm \rho\right) + 1\right]^{-1} dx$$

$$= k_B T \ln\left(1 + \exp\left(\eta \pm \rho\right)\right) \quad .$$

From this it follows that

$$\eta \pm \rho = \ln\left[\exp\left(\frac{\varepsilon_F\left(1 \pm \zeta\right)}{k_B T}\right) - 1\right] \quad ,$$

and hence for ρ

$$2\rho = \frac{2\zeta\left(\frac{k_B \Theta}{\varepsilon_F}\right)}{\left(\frac{k_B T}{\varepsilon_F}\right)}$$

$$= \ln\left[\frac{\exp\left(\frac{\varepsilon_F\left(1+\zeta\right)}{k_B T}\right) - 1}{\exp\left(\frac{\varepsilon_F\left(1-\zeta\right)}{k_B T}\right) - 1}\right] \quad .$$

With $\tau = \frac{k_B T}{\varepsilon_F}$ one obtains

$$\frac{k_B \Theta}{\varepsilon_F} = \frac{\tau}{2\zeta} \ln\left[\frac{\exp\left(\frac{1+\zeta}{\tau}\right) - 1}{\exp\left(\frac{1-\zeta}{\tau}\right) - 1}\right]$$

$$= \frac{1}{2} + \frac{\tau}{\zeta} \tanh^{-1}\left[\frac{\tanh\frac{\zeta}{2\tau}}{\tanh\frac{1}{2\tau}}\right] \quad . \tag{12.29}$$

The analytic form given in 12.29 is not very practical, especially if one wants to find an expression for $\zeta(T)$. But it can be used to derive some very interesting results:

1. Curie temperature: In the case $\zeta \to 0$, $\tau \to \tau_c = \frac{k_B T_c}{\varepsilon_F}$ one can easily determine the Curie temperature

$$\frac{\varepsilon_F}{k_B \Theta} = 1 - \exp\left(-\frac{1}{\tau_c}\right) \quad , \quad \frac{1}{\tau_c} = -\ln\left(1 - \frac{\varepsilon_F}{k_B \Theta}\right) \quad . \tag{12.30}$$

By chance the two equations given above are of exactly the same form as for the BCS theory of superconductivity. This is not completely surprising, since the formation of Cooper-pairs is also mediated by an interaction between the electrons forming these pairs. Within BCS theory, the electrons become coupled via a phonon. If magnetic order occurs, the electrons (spins) become coupled by the exchange interaction. Both phenomena describe the instability of the electron gas to a "many body" excitation, which lowers the total energy.

2. Susceptibility: For the susceptibility above T_c one assumes that for $\zeta = 0$, and $H \neq 0$ then

$$\frac{n\mu_B^2}{\chi} = \varepsilon_F \left[1 - \exp\left(-\frac{1}{\tau_c} \right) \right] - k_B \Theta \quad . \tag{12.31}$$

This result is analogous to (12.9).

3. Magnetization: For low temperatures one obtains a form similar to that given by (12.19)

$$\zeta = 1 - \tau \exp\left[-\frac{2}{\tau} \left(\frac{k_B \Theta}{\varepsilon_F} - 1 \right) \right] \quad . \tag{12.32}$$

For most systems the theoretical $\zeta(T)$ curves do not agree with experiment. The reason for this discrepancy is mainly that spinwaves and other collective excitations change the $\zeta(T)$ dependence considerably even at low temperature.

12.3 Weak Excitations with $\zeta \ll 1$

In the case of very weak itinerant systems one can expand the Stoner equations in powers of ζ and obtain to first order

$$\frac{2}{n} \mathcal{N}(\varepsilon_F)(k_B \Theta \zeta + \mu_B H) = \zeta \quad , \tag{12.33}$$

which is simply the susceptibility at $T = 0$. For finite temperatures one uses the temperature dependences of the two coefficients a and c from (3.18), (8.23), and (8.24)

$$\frac{2}{n} \mathcal{N}(\varepsilon_F)(k_B \Theta \zeta + \mu_B H) = \zeta(1 - aT^2) - \frac{1}{3}c\zeta^3 \quad . \tag{12.34}$$

For $c > 0$ and $a > 0$ and with (12.14) one finds

$$\frac{2}{n} \mathcal{N}(\varepsilon_F) k_B \Theta \zeta = 1 + \frac{1}{3}|c|\zeta^3 \quad ,$$

$$|a| T_c^2 = \frac{1}{3}|c|\zeta_0^3 \quad ,$$

$$\chi_0 = \frac{3\mu_B^2 \mathcal{N}(\varepsilon_F)}{|c|\zeta_0^2} \quad .$$

Introducing the magnetic moment $M = n\mu_B\zeta_0$ one obtains the equation for the magnetic isotherms of weak itinerant systems [133]

$$\left(\frac{M(H,T)}{M(0,0)} \right)^3 - \frac{M(H,T)}{M(0,0)} \left(1 - \frac{T^2}{T_c^2} \right) = \frac{2\chi_0 H}{M(0,0)} \quad . \tag{12.35}$$

Equation (12.35) also gives the Arrott plots which have been introduced earlier in Sect.5.5

$$M(H,T)^2 = M(0,0)^2 \left(1 - \frac{T^2}{T_c^2} \right) + \frac{2\chi_0 H}{M(H,T)} \quad . \tag{12.36}$$

One finds parallel lines with a constant slope which has the value $2\chi_0$. The intersection with the M^2axis is given by

$$M^2 = M(0,0)^2 \left(1 - \frac{T^2}{T_c^2}\right) \quad ,$$ (12.37)

so that the distance between the parallel lines is given by the temperature dependence of the magnetization. Linear Arrott plots are observed for all weakly ferromagnetic systems (ZrZn$_2$, Ni$_3$Al, Ni$_3$Ga, Ni-Pt, Fe-Ni, etc.). For other alloys and compounds there are characteristic deviations from linearity which are due to additional effects. For small fields one often finds deviations which are due to magnetization processes or to inhomogeneities of the sample. Deviations for large fields are often caused by spin fluctuations.

From (12.35) one can also determine the susceptibility below and above T_c. The inverse susceptibility is given by $\frac{dH}{dM}$. Rewriting (12.35) yields

$$\frac{M^3}{2\chi_0 M_0^2} - \frac{M}{2\chi_0}\left(1 - \frac{T^2}{T_c^2}\right) = H \quad ,$$

$$\frac{dH}{dM} = \frac{1}{\chi} = \frac{3M^2}{2\chi_0 M_0^2} - \frac{1}{2\chi_0}\left(1 - \frac{T^2}{T_c^2}\right) \quad .$$

One has two cases to consider:

1. $T > T_c$:

   Above the Curie temperature the magnetic moment becomes zero and one obtains

$$\chi = 2\chi_0 \left(\frac{T^2}{T_c^2} - 1\right)^{-1} \quad \text{for} \quad T > T_c \quad .$$ (12.38)

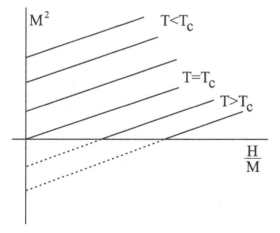

Fig. 12.4. Arrott plot of a weak itinerant magnet

2. $T < T_c$:

The case $T < T_c$ is a little bit more complicated since one has to consider the temperature dependence of the magnetization and make use of (12.37) to obtain

$$\chi = \chi_0 \left(1 - \frac{T^2}{T_c^2}\right)^{-1} \quad \text{for} \quad T < T_c \quad . \tag{12.39}$$

Apart from the mathematical formalism there must exist a physical reason for this surprising result that the susceptibility in the magnetic state is half of the susceptibility of the non-magnetic state. In the Stoner model there is no magnetic order and thus no molecular field above T_c. If one applies a magnetic field one creates spins from the "spin vacuum". Below the Curie temperature all spins are parallel to the molecular field and an applied field has to flip the spin from $-$ to $+$, which explains the factor 2. A classical analogue is the reflection of a ball from a solid wall. If one releases the ball from the hand it has the momentum p. Once the ball becomes reflected its momentum changes by $2p$.

To compare the susceptibility above T_c with a Curie–Weiss behavior one writes the susceptibility (12.38)

$$\frac{1}{\chi} = \frac{(T - T_c)(T + T_c)}{2\chi_0 T_c^2} \quad . \tag{12.40}$$

The first bracket would describe the Curie–Weiss law, but the whole expression shows a quadratic rather than a linear dependence. Calculating the Curie constant gives

$$
\begin{aligned}
C &= \frac{d\left(\frac{1}{\chi}\right)}{dT} \\
&= \frac{T}{\chi_0 T_c^2} \quad ,
\end{aligned}
\tag{12.41}
$$

which depends linearly on temperature rather than being a "constant".

At the end of this chapter one should recall that the Stoner model in general fails to describe the finite temperature properties of magnetic systems. The reason is that the excitations (single particle excitation also called Stoner excitations) which are thought to destroy magnetic order scale with T_F and are thus much too weak at normal temperature to create any sizeable effect. Since the Stoner exchange I_s (or Θ) is an atom specific quantity, the Stoner model only describes the conditions for the formation of a moment at a given atom. The interactions leading to long range order are only very indirectly accounted for via the change of the density of states upon alloying. It is therefore not surprising that the finite temperature Stoner model is today of merely historical interest.

13. Hubbard Model

Parallel to the development of band structure theory there was a search for "simple" toy-models to explain solid state magnetism. The Hubbard model combines electron hopping between neighboring sites and the Coulomb repulsion of electrons at the same site. With this onsite Coulomb repulsion it corrects for the usually neglected (or underestimated) electron correlation. Electron correlation means that electrons do not move independently of each other but of course feel their pairwise repulsion. However in most of the usual approximations to calculate the properties of the electron system, the influence of the other electrons is put into a mean field which has to be determined self-consistently. Within this approximation correlation effects are averaged out. The most prominent examples for this treatment are the Hartree- and the Hartree–Fock-method and all electronic structure methods based on them. These features are combined within the Hubbard Hamiltonian, [42, 135, 136] which in its simplest form reads

$$H = \sum_{ij\sigma} t_{ij} c_{i\sigma}^\dagger c_{j\sigma} + U \sum_i n_{i\uparrow} n_{i\downarrow} \quad . \tag{13.1}$$

The electron hopping is controlled by the t_{ij}, whereas the parameter U (the famous Hubbard-U) models the Coulomb interaction. $c_{i\sigma}^\dagger, c_{i\sigma}$ are the fermion creation and annihilation operators for an electron with spin σ on site i and $n_{i\sigma} = c_{i\sigma}^\dagger c_{i\sigma}$ is the related ladder operator counting the occupation on site i. These fermion operators obey the following anti-commutator rules

$$\left[c_{i\sigma}^\dagger, c_{j\sigma'} \right] = \delta_{ij}\delta_{\sigma\sigma'} \quad \left[c_{i\sigma}^\dagger, c_{j\sigma'}^\dagger \right] = [c_{i\sigma}, c_{j\sigma'}] = 0 \quad . \tag{13.2}$$

Depending on the sign of U the Hubbard Hamiltonian describes various phases:

- $U > 0$ (*repulsive*): paramagnetic metallic, ferromagnetic metallic, antiferromagnetic insulating,
- $U < 0$ (*attractive*): normal Fermi liquid, superconducting, charge density wave (insulator), normal Bose liquid (insulator).

U is however not the only parameter. There exists also the strength of the electron hopping t_{ij} and the temperature T. Further "hidden" variables are the dimensionality of the system and the structure of the crystal lattice.

Since the band filling (thus the number of electrons) plays an important role also the electron density $n = n_\uparrow + n_\downarrow$ can also be regarded as a parameter of the Hubbard Hamiltonian. Since the Pauli principle avoids that two electrons with like spin occupy the same lattice site one finds $n_\sigma \leq 1$ for $\sigma = \uparrow, \downarrow$ and thus $n \leq 2$. The special case $n = 1$ describes the half filled band (antiferromagnetism).

A straightforward solution can be found in the unperturbed case $U = 0$. The solution is found by the Fourier transform of the respective operator. In the time dependent form one obtains

$$i\hbar \dot{c}_{i\sigma} = -\sum_j t_{ij} c_{j\sigma} \quad \Rightarrow \quad c_{i\sigma} = \frac{1}{\sqrt{N}} \sum_k e^{ikR_i} c_{k\sigma} \quad , \tag{13.3}$$

$$i\hbar \dot{c}_{k\sigma}(t) = \epsilon_k^0 c_{k\sigma}(t) \quad \Rightarrow \quad c_{k\sigma}(t) = e^{-\frac{i}{\hbar}\epsilon_k^0 t} c_{k\sigma}(0) \quad , \tag{13.4}$$

which as a solution describes a single electronic band in the tight binding limit

$$\epsilon_k^0 = \epsilon_0 - \sum_{j \neq i} t_{ij} e^{ik(R_i - R_j)} \quad . \tag{13.5}$$

For a 2^d square lattice one finds $\epsilon_k^0 = -2t_{ij} (\cos(k_x a) + \cos(k_y a))$ leading to a band width $W = 4t_{ij}$.

In a similar way one treats the case for $U \neq 0$ in the Hartree–Fock approximation where the Hubbard interaction $n_{i\uparrow} n_{i\downarrow}$ is replaced by $\bar{n}_{i\uparrow} n_{i\downarrow} + n_{i\uparrow} \bar{n}_{i\downarrow} - \bar{n}_{i\uparrow} \bar{n}_{i\downarrow}$. The entities $\bar{n}_{i\sigma}$ are thermodynamical averages which have to be determined self consistently. The physical interpretation of the Hartree–Fock approximation is that fluctuations in the double occupancy $n_{i\uparrow} n_{i\downarrow}$ are suppressed. The respective time dependent Schrödinger equation reads

$$\left(i\hbar \frac{d}{dt} - \epsilon_0 - U\bar{n}_{i-\sigma} \right) c_{i\sigma} = -\sum_j t_{ij} c_{j\sigma} \quad . \tag{13.6}$$

The solutions are two bands split by the Hubbard interaction U

$$\epsilon_k^\uparrow = \epsilon_k^0 + \frac{1}{2} U (\bar{n} + m) \quad , \quad \epsilon_k^\downarrow = \epsilon_k^0 + \frac{1}{2} U (\bar{n} - m) \quad , \tag{13.7}$$

$$\text{with} \quad \bar{n}_\uparrow = \frac{1}{2}(\bar{n} + m) \quad , \quad \bar{n}\downarrow = \frac{1}{2}(\bar{n} - m) \quad . \tag{13.8}$$

The respective band splitting is thus given by $\Delta E = Um = \epsilon_k^\uparrow - \epsilon_k^\downarrow$, which can be directly compared to the result obtained from the Stoner model (8.11). In complete analogy to the Stoner model one also arrives at a criterion for magnetic ground state which is fulfilled if $UN(\varepsilon_F) \geq 1$ an expression which is in analogy to (8.15).

It is found that within the Hartree–Fock approximation the Hubbard model and the Stoner model become equivalent. The suppression of fluctuations in the double occupancy leads to an effective mean field treatment where Um represents the resulting self-consistent molecular field.

The missing physics is again the orientational disorder. This is well described by the Heisenberg model but hard to see for itinerant electrons where in the Hartree–Fock approximation the paramagnetic state becomes the non-magnetic state. However, what is required for the paramagnetic state is that the average over the magnetic moments vanishes

$$\left\langle \sum_i m_i \right\rangle = \left\langle \sum_i (n_{i\uparrow} - n_{i\downarrow}) \right\rangle \quad , \tag{13.9}$$

but the individual moment at site i remains, so that $m_i = n_{i\uparrow} - n_{i\downarrow} \neq 0$ for times τ such that $\frac{\hbar}{W} \leq \tau \leq \frac{\hbar}{k_B T}$.

13.1 Beyond Hartree–Fock

In general the Hubbard model hardly ever describes a ferromagnetic state. In most cases one finds an antiferromagnetic ground state. One of the few cases where a ferromagnetic ground state can be stabilized is the so called Nagaoka state [137]. In the limit of strong (infinite) Hubbard repulsion at half band filling one would describe an insulating state (Mott insulator [138, 139]). One can now ask the question about what will happen if one allows for a hole in the half filled band. For this scenario, Nagaoka found a ferromagnetic ground state for the simple cubic and the body centered cubic lattice. The conditions for the stabilization of a ferromagnetic state are extremely subtle, e.g. for close packed lattices (hcp or fcc) ferromagnetic ordering is unstable. In an attempt to go beyond the Hartree–Fock approximation which overemphasizes magnetism, Gutzwiller [140] proposed an approximation which promotes more competitive paramagnetic solutions. In the case of strong Hubbard repulsion $U >> t$ Hartree-Fock favors magnetic solutions, since these are the only (in the single particle picture) allowable type of correlations which ensure that the Hubbard repulsion is avoided. Gutzwiller's idea was to construct a wave function which reduces the probability of finding doubly occupied atoms without associated magnetic coherence. The Gutzwiller wave function reads

$$|\Psi_G\rangle = \prod_i [1 - \eta n_{i\uparrow} n_{i\downarrow}] |\psi_0\rangle = P(\eta) |\psi_0\rangle \quad , \tag{13.10}$$

where $|\psi_0\rangle$ is the wave function of the single particle ground state and η is a variational parameter which varies between 0 and 1. The operator $P(\eta)$ only effects components in real space which have doubly occupied atoms, and these components are reduced by a factor $(1 - \eta)^D$ where the exponent D is given by $D = \sum_i n_{i\uparrow} n_{i\downarrow}$ counting the doubly occupied sites. When $U \to \infty$ so $\eta \to 1$ so that all doubly occupied sites are projected out completely since $P(1)^2 = P(1)$. By applying the variational principle, η is determined such as to minimize the ground state energy calculated for Ψ_G. Unfortunately this variational problem cannot be solved analytically but there exists an approximation to the problem which is known as "Gutzwiller-approximation" which however becomes exact in the limiting case of infinite dimensionality. On the basis of numerical simulations for a two-dimensional square lattice it could be shown that while Hartree–Fock predicts ferromagnetism at any band filling, the Gutzwiller Ansatz indeed gives a paramagnetic ground state for low and high band fillings [141]. The Hubbard model with its various solutions provides a fascinating field of research. However, only during the last decade some connection between the "pure" Hubbard model and all-electron band structure calculations was made. Some pioneering work in this direction was done by W. Nolting and the reader is referred to [142] and references given therein.

14. Landau Theory for the Stoner Model

14.1 General Considerations

The Landau theory of phase transitions is a phenomenological description of the behavior of the free energy F as a function of an order parameter [143]. One can freely choose this order parameter as long as one obeys the inherent symmetry of the problem. This parameter can be temperature, volume, correlation length (superconductivity) or alloy concentration, etc. In the case of magnetic systems the role of the order parameter is usually taken by the magnetic moment M. The free energy for the magnetic system is then written as a polynomial in M and, by applying the rules of thermodynamics, its other physical properties can be derived. For a simple ferromagnet without an external field the free energy can be written

$$F = F_0 + a_2 M^2 + a_4 M^4 \quad . \tag{14.1}$$

The magnetic moment M enters only in even powers because only even terms are invariant under a reversal in the sign of magnetization (only for even powers of M is time reversal symmetry preserved). In lowest order the series can be truncated after the 4th order term, because if a_4 is taken to be positive, subsequent terms cannot alter the critical behavior of the system.

In Fig. 14.1 the variation of the Landau free energy for decreasing value of a_2 are shown (remember, a_4 is positive). The following four cases are depicted: (a) $a_2 > 0$, (b) $a_2 = 0$, (c) $a_2 \leq 0$, (d) $a_2 < 0$. The arrows denote the possible solutions for the equilibrium state. Due to inherent symmetry there always exists a solution for $+M$ and $-M$. The case $a_2 = 0$ corresponds to the critical temperature where spontaneous magnetization appears. One can thus write a_2 in the form

$$a_2 = a_2' t \quad \text{with} \quad t = \frac{T - T_c}{T_c} \quad . \tag{14.2}$$

With the variation of t one describes a continuous transition of the magnetization (however, its derivative changes discontinuously). This means that one can study the respective critical exponents (see Sect. 5.9). The easiest one to calculate is β. Since the equilibrium magnetization is given by the first derivative of the free energy, one obtains

$$\frac{dF}{dM} = 2a_2'tM + 4a_4M^3 = 0 \quad , \tag{14.3}$$

$$\Rightarrow M^2 = -\frac{a_2'}{2a_4} t \quad . \tag{14.4}$$

from which it follows immediately that for $t < 0$

$$M \propto (-t)^{\frac{1}{2}} \quad , \tag{14.5}$$

which is the mean field value $\beta_{mf} = \frac{1}{2}$. To calculate the critical exponent for the specific heat one uses (14.4) to rewrite the free energy

$$F = F_0 - \frac{a_2'}{4a_4} t^2 \quad . \tag{14.6}$$

The specific heat is given by $c_m = -T\frac{d^2F}{dT^2}$

$$c_m = \frac{a_2'}{2a_4T_c} (t+1) \quad , \tag{14.7}$$

which becomes constant if one approaches T_c ($t = 0$) from low temperatures. Coming from temperatures above always gives $c_m = 0$ because above T_c the magnetization M goes to zero and so does the free energy. The specific heat exhibits a discontinuity; its critical exponent is thus zero, $\alpha_{mf} = 0$.

To determine the critical exponent for the susceptibility γ (defined as $\chi \propto |t|^{-\gamma}$) one calculates the second derivative of the free energy and obtains

$$\frac{d^2F}{dM^2} = \frac{1}{\chi} = -4a_2't \quad , \tag{14.8}$$

$$\Rightarrow \gamma = 1 \quad , \tag{14.9}$$

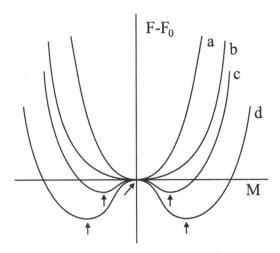

Fig. 14.1. Variation of the Landau free energy for decreasing values of a_2

which is again the mean field value. The exponents obtained from the Landau free energy are those for any mean-field model. Any model whose symmetry leads to a Landau expansion must have the same mean-field critical exponents. Ising-, XY-, and Heisenberg models are examples. A useful example is provided by the Ornstein–Zernicke extension to Landau theory which is discussed in Sect. F.

14.2 Application to the Stoner Model

One writes the free energy

$$F = \frac{1}{2}AM^2 + \frac{1}{4}BM^4 - MH \quad . \tag{14.10}$$

From the extremal conditions ($\frac{dF}{dM} = 0$ and $\frac{d^2F}{dM^2} > 0$) and using (12.39), the coefficients A and B can be derived easily

$$A = -\frac{1}{2\chi_0}\left(1 - \frac{T^2}{T_c^2}\right) \quad , \tag{14.11}$$

$$B = \frac{1}{2\chi_0 M_0^2} \quad . \tag{14.12}$$

Furthermore one finds the relations

$$M_0^2 = -\frac{A}{B} \quad , \tag{14.13}$$

$$\triangle F = -\frac{M_0^2}{8\chi_0} \quad , \tag{14.14}$$

$$M^2 = M_0^2\left(1 - \frac{T^2}{T_c^2}\right) \quad , \tag{14.15}$$

where M_0 is the equilibrium moment at $T = 0$, $\triangle F$ is the difference in energy between the magnetic and the non-magnetic state at $T = 0$, and M is the magnetic moment at a given temperature T. Replacing A, B, and M in (14.10) by the respective expressions given by (14.11), (14.12), and (14.15) yields the temperature dependence of the free energy, which for $H = 0$ reads

$$F_{\mathrm{m}} = -\frac{M_0^2}{8\chi_0}\left(1 - \frac{T^2}{T_c^2}\right)^2 \quad . \tag{14.16}$$

Equation (14.16) can now be used to calculate the magnetic contribution to the specific heat and the entropy. The specific heat of a Stoner system is given by

$$
\begin{aligned}
c_{\mathrm{m}} &= -T\frac{d^2F}{dT^2} \\
&= -\frac{M_0^2}{2\chi_0 T_c}\frac{T}{T_c}\left(1 - 3\frac{T^2}{T_c^2}\right) \quad .
\end{aligned} \tag{14.17}
$$

Since in the Stoner model the magnetic moment above T_c vanishes, c_m also becomes zero. One thus finds a pronounced discontinuity in the specific heat at T_c which has the value

$$\triangle c_m = \frac{M_0^2}{\chi_0 T_c} \tag{14.18}$$

For low temperature the behavior of c_m is determined by the linear term in (14.17). Formulating the specific heat in analogy to (2.47), one obtains

$$c_m = \gamma_m T \quad \text{with} \quad \gamma_m = -\frac{M_0^2}{2\chi_0 T_c^2} \quad , \tag{14.19}$$

where γ_m is often termed the "Wohlfarth correction". Equation(14.17) shows that the specific heat is proportional to T and T^3. Unfortunately at low T the specific heat of the electrons is also proportional to T and that of the phonons is proportional to T^3 which makes it very difficult to disentangle these contributions from the magnetic ones with any degree of certainty.

From the free energy one can also calculate the entropy via

$$\triangle S_m = \int_{T_1}^{T_2} \frac{c_m}{T} \mathrm{d}T \quad .$$

Taking the integral from $T = 0$ to $T = T_c$, $\triangle S_m$ vanishes. The complete functional dependence is

$$\triangle S_m = -\frac{M_0^2}{2\chi_0 T_c} \frac{T}{T_c} \left(1 - \frac{T^2}{T_c^2}\right) \quad . \tag{14.20}$$

The entropy S is zero for $T = 0K$ as required by the 3rd law of thermodynamics and it also vanishes for $T \geq T_c$ where magnetism disappears in

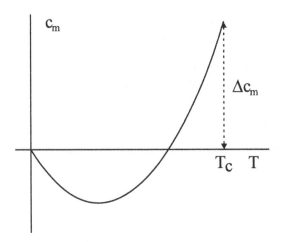

Fig. 14.2. Specific heat of a Stoner system

the Stoner model. In the intermediate temperature range a finite negative contribution is found.

To calculate magneto-elastic phenomena one introduces a volume dependence into (14.10). As the energy of an elastic system is proportional to V^2 (Hooke's law) one considers the square of the relative volume change ω^2. The interaction between volume and magnetic moment is described by the magneto–volume coupling constant C. How can the volume change due to the formation of a magnetic moment? Consider a slightly more than half filled d-band as in non-magnetic bcc Fe (Fig. 8.5) If the non-magnetic bands become split by the onset of magnetism electrons are removed from spin-down bonding states which now occupy spin-up antibonding states. This leads to a weakening of the bond strength and subsequently to a volume expansion. This phenomenon is also called volume magnetostriction. In general it is observed that the volume in the magnetically ordered state is always larger than in the non-magnetic state. With the new parameters the free energy reads

$$F = \frac{1}{2}AM^2 + \frac{1}{4}BM^4 - MH + \frac{1}{2\kappa}\omega^2 - C\omega\left(M(H,T)^2 - M(0,T)^2\right).$$

(14.21)

The new parameters in (14.21) are: κ, the compressibility; C, the magneto–volume coupling constant; and $\omega = (V - V_0)/V = \Delta V/V$, the relative volume change.

The relation between ω and P is given by $\omega = -\kappa P$.

The free energy now depends on two variables M and V. The equilibrium is given by the nodes of the magnetic equation of state $\frac{dF}{dM}\big|_V = 0$ and the mechanical equation of state $\frac{dF}{dV}\big|_M = 0$. From the latter one obtains

$$\omega = \kappa C\left(M(H,T)^2 - M(0,T)^2\right) \quad .$$

(14.22)

If one plots ω versus $M(H,T)^2$ one gets parallel lines. From the slope one obtains a value of the magneto-volume coupling constant C. From (14.21) one can also calculate the pressure dependence of the Curie temperature

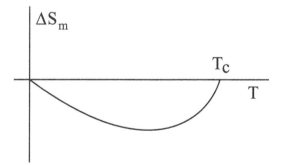

Fig. 14.3. Entropy of a Stoner system

$$\left(1 - \frac{T_c^2(P)}{T_c^2(0)}\right) = -4\chi_0 C\omega \quad ,$$

$$\Rightarrow T_c^2(P) = T_c^2(0)(1 - 4\chi_0 C\kappa P) \tag{14.23}$$

and the critical pressure for the disappearance of magnetism P_c

$$P_c^{-1} = 4\chi_0 C\kappa \quad , \tag{14.24}$$

$$\Rightarrow T_c^2(P) = T_c^2(0)\left(1 - \frac{P}{P_c}\right) \tag{14.25}$$

The pressure dependence given by (14.25) is observed very often. Equation (14.25) yields a general form which contains only ground state quantities (besides T_c which is just a scaling parameter). An experimental determination of P_c is often impossible because the values can become very large. It is easier to measure the slope $\frac{dT_c}{dP}$ which is given by

$$2T_c\frac{dT_c}{dP} = -4T_c^2\chi_0 C\kappa \quad ,$$

$$\Rightarrow \frac{dT_c}{dP} = -2T_c\chi_0 C\kappa \quad . \tag{14.26}$$

With: $\chi_0 = \mu_B^2 \mathcal{N}(\varepsilon_F)\frac{T_F^2}{T_c^2}$ (14.26) becomes

$$\frac{dT_c}{dP} = -\frac{\alpha}{T_c} \quad \text{with} \quad \alpha = 2\kappa C\mu_B^2 \mathcal{N}(\varepsilon_F) T_F^2 \quad . \tag{14.27}$$

This result is in fair agreement with experiment. Many weak itinerant systems show this type of hyperbolic dependence (Fe-Ni Invar, Fe-Pt, Fe-Pd). In Table 14.1 below a few values are given for α and α/T_F^2

Table 14.1. Pressure coefficient α and pressure scaling factor α/T_F^2 for some weak itinerant ferromagnets

	α (K^2/kbar)	α/T_F^2 (kbar)
ZrZn$_2$	40 ± 4	5.9
Ni$_3$Al	28 ± 8	2.2
Ni-Pt	36 ± 2	9.0
Fe-Ni Invar	2000 ± 400	5

The thermal expansion coefficient α_m can also be calculated from (14.21)

$$\alpha_m = \frac{d\omega}{dT}$$

$$= -T\frac{2\kappa C M (0,0)^2}{T_c^2} \quad . \tag{14.28}$$

As for the specific heat, a discontinuity occurs at T_c with the value

$$\triangle\alpha_m = -\frac{2\kappa C M (0,0)^2}{T_c} \quad . \tag{14.29}$$

Comparing (14.18) for $\triangle c_m$, (14.29) for $\triangle\alpha_m$, and (14.27) for $\frac{dT_c}{dP}$ one can apply a general thermodynamical relation

$$\frac{\triangle\alpha_m}{\triangle c_m} = \frac{d\ln T_c}{dP} = -2\kappa C\chi_0 \quad . \tag{14.30}$$

Equation (14.30) is a so called Ehrenfest relation. The Ehrenfest relations are model independent relations for the behavior of thermodynamical variables at critical points.

Another important quantity is the bulk modulus $\mathcal{B}$. It is defined as the hypothetical pressure necessary to reduce the volume by a factor $1/2$. Thermodynamically it is given by

$$\mathcal{B} = V(0,0)\frac{d^2 F}{dV^2} \quad . \tag{14.31}$$

Also $\mathcal{B}$ shows a discontinuity at T_c which has the value

$$\triangle\mathcal{B} = 4\chi_0 M (0,0)^2 C^2 \tag{14.32}$$

Equation (14.32) yields a direct possibility of determining the magneto–volume coupling constant from experiment. It should be noted once again that the discontinuities derived above for the Stoner model are in general too large as compared to experiment. The reason is of course that in the Stoner model the state above the Curie temperature is the true non-magnetic state where no magnetic moments whatsoever exist. In reality the true paramagnetic state is a state where local moments still exist but where the long range order has broken down. This difference in the description of the non-magnetic state is that reason why Stoner theory fails at finite temperature.

15. Coupling Between Itinerant and Localized Moments

The alloys and compounds formed between rare earth elements and 3d elements are of great technological importance. One example among many is the intermetallic compound $Nd_2Fe_{14}B$ which is a high performance permanent magnet. It is characterized by the largest energy product $BH_{max} = 360$ kJ/mol of all known permanent magnet materials. For a rectangular hysteresis, the energy product BH_{max} measures the area under the hysteresis curve defined as the product of the remanent induction B_r times the coercivity H_c. Unfortunately, due to its Curie temperature of only 600K, $Nd_2Fe_{14}B$ is restricted to applications only around ambient temperatures. The reason for this limitation is the strong temperature dependence of the demagnetization which even at 400K makes $Nd_2Fe_{14}B$ inferior to the older $SmCo_5$ magnets. It is thus of vital importance to study the mechanism which determines the ordering temperature in such systems. In the case of $Nd_2Fe_{14}B$ and also of $SmCo_5$ the rare earth atom with its large localized f-electron moment not only produces a strong anisotropy (which is important for a large energy product) but also couples to the 3d transition metal atoms which are the major carriers of the magnetization. A review of the experimental results of this class of materials can be found in [144].

A elegant way of formulating the coupling between localized and initerant moment was given by Bloch et al. [145]. The localized magnetic moment M_{loc} is proportional to the angular momentum J and is given by the relation

$$M_{loc} = \mu_B g_J J \quad . \tag{15.1}$$

The localized moment should couple to the spins S of the itinerant electrons via

$$2K (g_J - 1) JS \quad . \tag{15.2}$$

In (15.1) and (15.2), J is the total angular momentum of the rare earth atom, g_J is the Landé factor, and K is a coupling constant. The Hamiltonian for the coupled systems with an external field H_{ext}^z applied in the z direction is given by

$$\mathcal{H} = \mathcal{H}_{band} + [2K (g_J - 1) \langle S_z \rangle + \mu_B g_J H_{ext}^z] \sum_i J_i^z + 2N\mu_B \langle S_z \rangle H_{ext}^z \quad , \tag{15.3}$$

where $\mathcal{H}_{\text{band}}$ is the contribution from the non-magnetic electronic band structure, $\langle S_z \rangle$ the average spin in the 3d band per molecular unit that is per rare earth atom, and N is the number of rare earth atoms. The magnetization of the 3d band is given by

$$M_{3d} = -2N\mu_B \langle S_z \rangle \quad . \tag{15.4}$$

From (15.3) one can determine the field H_K which acts on the localized rare earth atoms and represents the influence of the itinerant 3d spins via the coupling parameter K

$$H_K = H_{\text{ext}}^z - \frac{(g_J - 1)}{N g_J \mu_B^2} K M_{3d} \quad . \tag{15.5}$$

The total free energy contains the contribution of the 3d atoms $F_{3d}(M_{3d}, T)$, the contribution of the external field $-H_{\text{ext}}^z M_{3d}$ and the usual term from the Weiss model (Chap. 6) for an angular momentum J in a "molecular field" H_K

$$F = F_{3d}(M_{3d}, T) - H_{\text{ext}}^z M_{3d} - N k_B T \ln \frac{\sinh\left(\left(J + \frac{1}{2}\right) y\right)}{\sinh\left(\frac{y}{2}\right)} \quad , \tag{15.6}$$

with

$$y = \frac{\mu_B g_J H_K}{k_B T} \quad . \tag{15.7}$$

From the equilibrium condition $dF/dM_{3d} = 0$ one obtains

$$-\frac{dF_{3d}(M_{3d}, T)}{dM_{3d}} = H_{\text{ext}}^z - \frac{g_J - 1}{\mu_B} K J B_J(J, y) \quad , \tag{15.8}$$

where the Brillouin function $B_J(J, y)$ (6.8) is defined as

$$B_J(J, y) = \frac{2J + 1}{2J} \coth\left(\left(J + \frac{1}{2}\right) y\right) - \frac{1}{2J} \coth\left(\frac{y}{2}\right) \quad . \tag{15.9}$$

Applying the usual high temperature approximation (6.10) of $B_J(J, y)$

$$B_J(J, y) \simeq \frac{1}{3} y (J + 1) \tag{15.10}$$

and for zero external field, $H_{\text{ext}}^z = 0$, one obtains

$$\frac{dF_{3d}(M_{3d}, T)}{dM_{3d}} = \frac{(g_J - 1)^2}{N\mu_B^2} K^2 \frac{J(J + 1)}{3k_B T} M_{3d} \quad . \tag{15.11}$$

Let us assume that the free energy of the 3d atoms can be written in form of a Landau type expansion (14.10)

$$F_{3d}(M_{3d}, T) = \frac{A(T)}{2} M_{3d}^2 + \frac{B(T)}{4} M_{3d}^4 + \frac{C(T)}{6} M_{3d}^6 + \dots \tag{15.12}$$

The rhs of (15.11) renormalizes the coefficient $A(T)$ of the free energy $F_{3d}(M_{3d}, T)$ so that the equilibrium condition reads

$$0 = \frac{dF}{dM_{3d}} = M_{3d}\left(A\left(T\right) + \frac{\left(g_J - 1\right)^2}{N\mu_B^2}K^2\frac{J\left(J+1\right)}{3k_BT}\right) + \ldots \qquad (15.13)$$

The coefficient $A\left(T\right)$ is given by $-1/\left(2\chi_{3d}\left(T\right)\right)$. The inverse susceptibility of the coupled system is thus given by

$$\chi^{-1} = \frac{d2F}{dM_{3d}^2} = \left(-\frac{1}{2\chi_{3d}\left(T\right)} + \frac{\left(g_J - 1\right)^2}{N\mu_B^2}K^2\frac{J\left(J+1\right)}{3k_BT}\right) + \ldots \qquad (15.14)$$

At the Curie temperature T_c the susceptibility diverges so that $\chi^{-1} = 0$ which yields a condition for the Curie temperature

$$k_BT_c = \frac{\left(g_J - 1\right)^2}{N\mu_B^2}K^2\frac{2J\left(J+1\right)}{3}\chi_{3d}\left(T_c\right) \quad . \qquad (15.15)$$

Equation (15.15) is a very general expression for the Curie temperature of coupled systems. It can be applied to systems where the 3d atoms carry a magnetic moment by themselves which of course becomes changed by the presence of the field exerted by the localized moment, but also to alloys where the 3d atoms are genuinely non-magnetic but become polarized by the localized moment. The coupling constant K can be determined from (15.8) by setting $T = 0K$ so that $B_J\left(J, y\right) = 1$ which gives

$$K = \frac{\mu_B}{J\left(g_J - 1\right)}\frac{dF_{3d}\left(M_{3d}, T = 0\right)}{dM_{3d}} = \frac{\mu_B}{J\left(g_J - 1\right)}H_{3d} \quad . \qquad (15.16)$$

Replacing the field H_{3d} by M_{3d}/χ_{3d} and by applying (15.1) one finds

$$K = \frac{g_j\mu_B^2}{\left(g_J - 1\right)}\chi_{3d}^{-1}\frac{M_{3d}}{M_{loc}} \quad . \qquad (15.17)$$

The coupling constant, which is of the dimension of an energy, describes a linear relation between the 3d moment and the localized moment, which is in agreement with the physical intuition.

The model introduced here is also known under the names $s - d$ model or $d - f$ model, depending which type of atom is supposed to provide the itinerant and which the localized moment. In the original paper [145] it was applied to the problem of the order of the phase transition in ACo_2 compounds, where A=Er, Ho, Dy, Tb, Gd. A different application was found for the ordering temperature of transition metal impurities in a Pd host lattice [161] where a combination with spin fluctuation theory was formulated.

16. Origin of the Molecular Field

Starting from the early days of Weiss, most models postulated an inner field to explain the formation of magnetic order. Historically this field is called molecular field. Since a classical formulation (theorem of Bohr and van Leeuwen; Chap. 1) cannot explain magnetic order, it already became clear in the 1920s that the interactions leading to solid state magnetism must be of quantum mechanical nature.

16.1 Heitler–London Theory for the Exchange Field

The origin of the molecular field is the interaction between the electrons. These are electrostatic effects of coulomb type. Very early on Heitler and London [21] gave an LCAO–type solution for the H_2 molecule. There they also took into consideration the spin of the electrons which means that the wavefunction must be antisymmetric under the exchange of two electrons. However, because of the atomic–like wavefunctions used in their model, the Heitler–London solution is only valid for localized and not for itinerant electrons. Nevertheless it gives a direct explanation of the nature of the interaction incorporated into a molecular field in terms of the exchange interaction.

To write down the respective Schrödinger equation for the H_2-molecule one considers the geometry given in Fig. 16.1

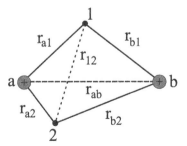

Fig. 16.1. Schematic geometry of the H_2 molecule. The grey spheres represent the protons and the black spheres the electrons. $\boldsymbol{r}_{xy}$ are the respective distances

$$\left(\frac{\hbar^2}{2m_e} (\nabla_1^2 + \nabla_2^2) + U_m - \left[\frac{e^2}{r_{ab}} - \frac{e^2}{r_{a1}} - \frac{e^2}{r_{a2}} - \frac{e^2}{r_{b1}} - \frac{e^2}{r_{b2}} + \frac{e^2}{r_{12}} \right] \right) \Psi_m = 0.$$
(16.1)

U_m is the exact energy eigenvalue for the molecular wavefunction Ψ_m. For two isolated H-atoms the Schrödinger equations are

$$\frac{\hbar^2}{2m_e} \nabla^2 \psi_{a1} + (E_0 + \frac{e^2}{r_{a1}}) \psi_{a1} = 0 \quad ,$$
(16.2)

$$\frac{\hbar^2}{2m_e} \nabla^2 \psi_{b2} + (E_0 + \frac{e^2}{r_{b2}}) \psi_{b2} = 0 \quad .$$
(16.3)

E_0 is the eigenvalue of the isolated atom, ψ_{a1} and ψ_{b1} are the atomic wave functions (e.g. $1s$-functions). Without the electron–electron interaction, the ground state of the H_2 molecule would be 4-fold degenerate: There are two electrons and each can have $+$ or $-$ spin so one is involved with

$$\psi_a \alpha \ , \ \psi_a \beta \ , \ \psi_b \alpha \ , \ \psi_b \beta \quad \text{with} \quad \alpha, \beta = +\frac{1}{2}, -\frac{1}{2} \quad .$$

The total spin S can take two values: $S = 0$, $\mathcal{M}_S = 0$ and $S = 1$, $\mathcal{M}_S = -1, 0, +1$.

The molecular wavefunction Ψ_m cannot be calculated analytically. The usual way to circumvent this ever-present problem is to approximate the "unsolvable" many-body problem by using linear combinations of known wavefunctions. To set up these linear combinations one introduces linear combination coefficients which are then used as variational parameters to minimize the total energy of this approximate solution. To solve the Schrödinger equation of the H_2 molecule one uses exactly this method.

Distributing the electrons among these four wavefunctions one can construct four determinant wavefunctions. The use of determinant wavefunctions (Slater determinant) satisfies the antisymmetry of the problem, because exchange of two electrons means exchange of two rows of the determinant and consequently a change of sign. These four determinant wavefunctions are

$$\Psi_1 = \begin{vmatrix} \psi_a \alpha (1) & \psi_b \alpha (1) \\ \psi_a \alpha (2) & \psi_b \alpha (2) \end{vmatrix} \Rightarrow \mathcal{M}_S = 1 \quad ,$$
(16.4)

$$\Psi_2 = \begin{vmatrix} \psi_a \beta (1) & \psi_b \alpha (1) \\ \psi_a \beta (2) & \psi_b \alpha (2) \end{vmatrix} \Rightarrow \mathcal{M}_S = 0 \quad ,$$
(16.5)

$$\Psi_3 = \begin{vmatrix} \psi_a \alpha (1) & \psi_b \beta (1) \\ \psi_a \alpha (2) & \psi_b \beta (2) \end{vmatrix} \Rightarrow \mathcal{M}_S = 0 \quad ,$$
(16.6)

$$\Psi_4 = \begin{vmatrix} \psi_a \beta (1) & \psi_b \beta (1) \\ \psi_a \beta (2) & \psi_b \beta (2) \end{vmatrix} \Rightarrow \mathcal{M}_S = -1 \quad .$$
(16.7)

Here for instance $\psi_a \alpha (1)$ means: electron 1 at atom a with spin $+1/2$. Without interaction, these wavefunctions lead to the eigenvalue

$$U_0 = 2E_0 \quad . \tag{16.8}$$

When the interaction is switched on, the fourfold degenerate state splits into a singlet ($\mathcal{S} = 0$, $\mathcal{M}_\mathcal{S} = 0$) and a triplet ($\mathcal{S} = 1$, $\mathcal{M}_\mathcal{S} = -1, 0, +1$) state where a priori either of the two could be the ground state.

$$2E_0 \quad \frac{\mathcal{S} = 1, \mathcal{M}_\mathcal{S} = -1, 0, +1}{\mathcal{S} = 0, \mathcal{M}_\mathcal{S} = 0} \quad ,$$

$$2E_0 \quad \frac{\mathcal{S} = 0, \mathcal{M}_\mathcal{S} = 0}{\mathcal{S} = 1, \mathcal{M}_\mathcal{S} = -1, 0, +1}$$

The aim is now to find out which interaction is responsible for making the "antiferromagnetic" singlet ($\mathcal{S} = 0$ the spins are antiparallel) or the ferromagnetic triplet ($\mathcal{S} = 1$ the spins are parallel) lower in energy.

Since one considers isolated atoms one can directly formulate an orthogonality relation between the electron wavefunctions

$$S_{\alpha\beta} = \int \psi_a \alpha \psi_b \beta d\tau = 0 \quad , \tag{16.9}$$

$$S_{\alpha\alpha} = \int \psi_a \alpha \psi_b \alpha d\tau = S_{\beta\beta} = \int \psi_a \beta \psi_b \beta d\tau = S \quad . \tag{16.10}$$

In our case the value of the overlap integral S is directly related to the orthogonality relations. Two cases are possible

$$r_{ab} = 0 \Rightarrow S = 1 \quad \text{the wavefunctions are identical,}$$

$$r_{ab} \to \infty \Rightarrow S = 0 \quad \text{no overlap.}$$

One constructs the approximation for the molecular wavefunction as a sum over all four determinant wavefunctions with variational parameters c_i

$$\Psi = \sum_{i=1}^{4} c_i \Psi_i \tag{16.11}$$

To perform the calculation a number of abbreviations are introduced, starting with

$$\mathcal{H}\Psi = U\Psi \quad ,$$

$$\text{with} \quad \mathcal{H} \doteq 2E_0 + V_0 \quad ,$$

$$\text{and} \quad V_0 = e^2 \left[\frac{1}{r_{ab}} - \frac{1}{r_{a2}} - \frac{1}{r_{b1}} + \frac{1}{r_{12}} \right] \quad ,$$

where U is the eigenvalue to the approximate wavefunction Ψ. It is also convenient to introduce the following spin functions

$$\sigma_1 = \alpha\,(1)\,\alpha\,(2) \quad \mathcal{M}_S = +1 \quad,$$
$$\sigma_2 = \beta\,(1)\,\alpha\,(2) \quad \mathcal{M}_S = 0 \quad,$$
$$\sigma_3 = \alpha\,(1)\,\beta\,(2) \quad \mathcal{M}_S = 0 \quad,$$
$$\sigma_4 = \beta\,(1)\,\beta\,(2) \quad \mathcal{M}_S = -1 \quad.$$

Rewriting the Schrödinger equation (16.1) now yields

$$(U - 2E_0)\,\Psi - V_0\Psi = 0 \quad. \tag{16.12}$$

Multiplying (16.12) with the product $\psi_a\alpha\,(1)\,\psi_b\beta\,(2)$ form the rhs gives

$$(U - 2E_0) \int_\sigma \alpha\,(1)\,\beta\,(2)\,\mathrm{d}\sigma \int_\tau \Psi\psi_a\,(1)\,\psi_b\,(2)\,\mathrm{d}\tau$$
$$= \int_\sigma \alpha\,(1)\,\beta\,(2)\,\mathrm{d}\sigma \int_\tau \Psi V_0\psi_a\,(1)\,\psi_b\,(2)\,\mathrm{d}\tau \quad,$$

or in general

$$(U - 2E_0) \int_\sigma \sigma_k\mathrm{d}\sigma \int_\tau \Psi\psi_a\,(1)\,\psi_b\,(2)\,\mathrm{d}\tau = \int_\sigma \sigma_k\mathrm{d}\sigma \int_\tau \Psi V_0\psi_a\,(1)\,\psi_b\,(2)\,\mathrm{d}\tau.$$
$$\tag{16.13}$$

One can now replace Ψ for the sum in (16.11) and obtain

$$\sum_{i=1}^{4} c_i\,[H_{ik} + (2E_0 - U)\,S_{ik}] = 0 \quad, \tag{16.14}$$

where

$$S_{ik} = \int_\sigma \sigma_k\mathrm{d}\sigma \int_\tau \Psi_i\psi_a\,(1)\,\psi_b\,(2)\,\mathrm{d}\tau \quad, \tag{16.15}$$

and

$$H_{ik} = \int_\sigma \sigma_k\mathrm{d}\sigma \int_\tau \Psi_i V_0\psi_a\,(1)\,\psi_b\,(2)\,\mathrm{d}\tau \quad. \tag{16.16}$$

The variation of the total energy with respect to the coefficients c_i leads to a secular determinant of the form

$$\begin{vmatrix} H_{11} + ES_{11} & 0 & 0 & 0 \\ 0 & H_{22} + ES_{22} & H_{23} + ES_{23} & 0 \\ 0 & H_{32} + ES_{32} & H_{33} + ES_{33} & 0 \\ 0 & 0 & 0 & H_{44} + ES_{44} \end{vmatrix} = 0 \quad, \tag{16.17}$$

with $E = 2E_0 - U$.

The secular matrix is of block diagonal form. This means that one only has to diagonalize the block in the center. The first element can immediately be calculated: $S_{11} \to \mathcal{M}_S = +1$, $i = 1$, $k = 1$

$$\Psi_1 = \sigma_1 \left[\psi_a (1) \psi_b (2) - \psi_a (2) \psi_b (1) \right] \quad ,$$

$$S_{11} = \underbrace{\underbrace{\int_\sigma \sigma_1^2 d\sigma}_{=1} \int_\tau \psi_a (1) \psi_b (2) \psi_a (1) \psi_b (2) d\tau}_{=1}$$

$$- \underbrace{\underbrace{\int_\sigma \sigma_1^2 d\sigma}_{=1} \int_\tau \psi_a (2) \psi_b (1) \psi_a (1) \psi_b (2) d\tau}_{=S^2} \quad ,$$

$$S_{11} = 1 - S^2 \quad , \tag{16.18}$$

$$H_{11} = \underbrace{\underbrace{\int_\sigma \sigma_1^2 d\sigma}_{=1} \int_\tau |\psi_a (1)|^2 |\psi_b (2)|^2 V_0 d\tau}_{=C}$$

$$- \underbrace{\underbrace{\int_\sigma \sigma_1^2 d\sigma}_{=1} \int_\tau \psi_a (2) \psi_b (1) \psi_a (1) \psi_b (2) V_0 d\tau}_{=J} \quad ,$$

$$H_{11} = C - J \quad . \tag{16.19}$$

Here S is the overlap-, C the coulomb-, and J the exchange integral. The first matrix element is now given by

$$H_{11} + (2E_0 - U) S_{11} = 0 \quad ,$$

$$\Rightarrow U = U_1 = 2E_0 + \frac{C - J}{1 - S^2} \quad , \quad \mathcal{M}_S = +1 \tag{16.20}$$

The energy for H_{44} is calculated analogously and has the same value, but for $\mathcal{M}_S = -1$. The terms U_1 and U_4 describe states where both spins are parallel which leads to a "ferromagnetic" state with $\mathcal{S} = 1$.

Diagonalizing the block in the center of the determinant one can use the following relations which are due to the fact that the secular matrix is hermitian

$$S_{22} = S_{33} = \int_\sigma \sigma_2^2 d\sigma \int_\tau |\psi_a (1)|^2 |\psi_b (2)|^2 d\tau = 1 \quad , \tag{16.21}$$

$$H_{22} = H_{33} = \int_\sigma \sigma_2^2 d\sigma \int_\tau |\psi_a (1)|^2 |\psi_b (2)|^2 V_0 d\tau = C \quad , \tag{16.22}$$

$$S_{23} = S_{32} = \int_\sigma \sigma_3^2 d\sigma \int_\tau \psi_a (2) \psi_b (1) \psi_a (1) \psi_b (2) d\tau = S^2 \quad , \tag{16.23}$$

$$H_{23} = H_{32} = \int_\sigma \sigma_3^2 d\sigma \int_\tau \psi_a(2)\,\psi_b(1)\,\psi_a(1)\,\psi_b(2)\,V_0 d\tau = J \quad , \quad (16.24)$$

so that this block reads

$$\begin{vmatrix} C + (2E_0 - U) & J + (2E_0 - U)\,S^2 \\ J + (2E_0 - U)\,S^2 & C + (2E_0 - U) \end{vmatrix} = 0 \quad ,$$

which leads to the eigenvalues

$$U_2 = 2E_0 + \frac{C+J}{1+S^2} \quad , \quad \mathcal{M}_S = 0 \quad , \quad S = 0 \quad , \tag{16.25}$$

and

$$U_3 = 2E_0 + \frac{C-J}{1-S^2} \quad , \quad \mathcal{M}_S = 0 \quad , \quad S = 1 \quad . \tag{16.26}$$

The Heitler–London model leads indeed to a threefold degenerate state with parallel spin (triplet state) and a non-degenerate state with antiparallel spin (singlet state). If one now postulates that the "ferromagnetic" state ($S = 1$) should be lower in energy than the "antiferromagnetic" state ($S = 0$) one finds the relation

$$U(S=1) - U(S=0) = \frac{2\left(CS^2 - J\right)}{1 - S^4} < 0 \quad ,$$

and if the overlap can be neglected, one finds the well known result that the exchange interaction must be positive

$$J > 0 \quad . \tag{16.27}$$

The state of the H_2 molecule thus depends on the sign of the exchange integral. The magnitude of the exchange energy is about $2J$ and is proportional to the quantity which is called the molecular field. The exchange interaction is entirely of quantum mechanical origin and is an interaction of coulomb type. This is the reason why the values for the molecular field came out so unphysically large (Table 6.1). It must be stressed that this example only yields a plausible explanation and may not be applied to any itinerant system and to systems with more than one valence electron. However, there are a number of conclusions that can be drawn from it.

16.1.1 Magnetism of a Spin Cluster

One can try to generalize our result to a cluster of spins (cluster of atoms with spin $+$ or $-$).

$$\begin{array}{ccc} \downarrow & \downarrow & \uparrow \\ \uparrow & \uparrow & \downarrow \\ \uparrow & \downarrow & \uparrow \end{array}$$

One assumes that the central atom has z neighbors, of which x neighbors have spin up $(+1/2)$ and y neighbors have spin down $(-1/2)$, $x + y = z$. For each pair of atoms with parallel spin, the energy is

$$U = 2E_0 + \frac{C - J}{1 - S^2} \simeq 2E_0 + C - J \quad ,$$

and for each pair with antiparallel spin

$$U = 2E_0 + \frac{C + J}{1 + S^2} \simeq 2E_0 + C + J \quad .$$

Apart from spin independent constants one writes the total energy of the system

$$\Delta U = -xJ + yJ = -J(x - y) \quad .$$

Depending on whether the central atom has spin-up or spin-down one finds

$$\Delta U = \pm \frac{1}{2} J(x - y) \quad .$$

Introducing the relative magnetization $\zeta = \frac{(x+y)}{(x-y)}$ one obtains for ΔU

$$\Delta U = \pm \frac{1}{2} z J \zeta \quad .$$

To calculate the temperature dependence one uses classical statistics (the particles are on fixed lattice sites and thus distinguishable) and obtains

$$\zeta = \frac{\exp\left(\frac{1}{2} z \frac{J}{k_B T} \zeta\right) - \exp\left(-\frac{1}{2} z \frac{J}{k_B T} \zeta\right)}{\exp\left(\frac{1}{2} z \frac{J}{k_B T} \zeta\right) + \exp\left(-\frac{1}{2} z \frac{J}{k_B T} \zeta\right)} \tag{16.28}$$

$$= \tanh\left(\frac{1}{2} z \frac{J}{k_B T} \zeta\right) \quad .$$

This is of course the same result (D.4) as in Sect. D.. Comparing this equation with the result one obtained from the Weiss' molecular field model (6.18)

$$\text{Equation (6.18)} \rightarrow \zeta = \tanh\left(\frac{\zeta T_c}{T}\right) \quad ,$$

one finds a relation between the value of the exchange integral and the Curie temperature

$$\frac{k_B T_c}{J} = \frac{z}{2} \quad . \tag{16.29}$$

With this simple model one finds surprisingly good results for the value of the exchange integral which are of the order of

$$\frac{2 k_B T_c}{z} \approx 20 \text{ meV} \, .$$

The generalization of the Heitler–London model for the H_2 molecule to "solids" is only valid as long as the electrons are localized (compare with the Weiss model, Chap. 6). In the present form it is also restricted to systems which have only one valence electron. The model leads to essentially wrong results for the conductivity (ferromagnetic conductors and antiferromagnetic insulators).

16.1.2 Spinwaves for Localized Electrons

In a similar manner to that of the preceding section one now considers a linear chain of spins. For this many–particle system there exists a wave function Ψ and an energy eigenvalue U which satisfy a Schrödinger equation

$$\mathcal{H}\Psi = U\Psi \quad .$$

The aim is now to calculate the excitations of the spin system at low temperatures. One assumes that a number r of the N spins forming the chain are antiparallel to the others. N is not only the total number of spins, but also the total number of electrons in the system. Usually r will be much smaller than N and the magnetic moment is given by

$$M = \mu_B \left(N - 2r\right) \quad , \tag{16.30}$$

$$\zeta = \frac{M}{N\mu_B} = 1 - \frac{2r}{N} \quad .$$

One now enumerates the lattice sites such that the antiparallel spins occupy the positions $n \equiv n_1, n_2, n_3, \ldots, n_r$. Each electron on its lattice site n is described via an "atomic" wavefunction $\psi\left(n\right)$. The wavefunction of the spin system, in analogy to (16.11), is written as a linear combination of the individual determinant wavefunctions

$$\Psi = \sum_n \alpha\left(n\right) \Psi\left(n\right) \quad , \tag{16.31}$$

where the $\alpha\left(n\right)$ are the variational parameters. For a number n of reversed spins one can construct 2^n configurations. One now writes the Schrödinger equation, analogously to (16.14) in the form

$$\sum_n \alpha\left(n\right) \left[V\left(n, n'\right) - U\triangle\left(n, n'\right)\right] = 0 \quad , \tag{16.32}$$

with

$$\triangle\left(n, n'\right) = \int \Psi\left(n\right) \Psi\left(n'\right) \mathrm{d}\tau \quad , \tag{16.33}$$

$$V\left(n, n'\right) = \int \Psi\left(n\right) \mathcal{H}\Psi\left(n'\right) \mathrm{d}\tau \quad . \tag{16.34}$$

One can now calculate the matrix elements:

For an interaction leading to a diagonal element two neighboring parallel spins change sites. Obviously the distribution is unchanged. Off-diagonal elements are created when two neighboring antiparallel spins change sites and hence the distribution changes.

Diagonal Elements: The distribution n is changed to n', but where n' is equal to n

$$
\begin{array}{cccccc}
\uparrow & \uparrow & \rightleftharpoons \uparrow & \downarrow & \uparrow & n \\
\uparrow & \uparrow & \uparrow & \downarrow & \uparrow & n = n'
\end{array}
$$

$V(n, n')$ is thus given by (16.20) (as for H_{11})

$$
V(n, n') = NE_0 + NC - mJ \quad , \quad S = 1 \quad . \tag{16.35}
$$

Here m is the number of pairs of neighboring parallel spins. If m' is the number of pairs of neighbors with antiparallel spins one gets

$$
Nz = m + m' \quad , \quad z \ldots \text{number of nearest neighbors.} \tag{16.36}
$$

Off-diagonal Elements: If two neighboring antiparallel spins change sites, the distribution n is changed to the distribution n' and $n \neq n'$

$$
\begin{array}{cccccc}
\uparrow & \uparrow & \uparrow & \rightleftharpoons \downarrow & \uparrow & n \\
\uparrow & \uparrow & \downarrow & \uparrow & \uparrow & n \neq n'
\end{array}
$$

In this case $V(n, n')$ is given by H_{23} (16.24) and has the value

$$
V(n, n') = -J \quad . \tag{16.37}
$$

For $\triangle(n, n')$ one can use the orthogonality between two distributions n and n'

$$
\triangle(n, n') = \delta_{n, n'} \quad . \tag{16.38}
$$

For a distribution with r antiparallel spins the total energy is thus given by

$$
U\alpha(n) = (NE_0 + NC - mJ)\,\alpha(n) - J \sum_{n'} \alpha(n') \quad . \tag{16.39}
$$

By using the relation $m = Nz - m' = Nz - \sum_{n'} 1$ (16.36) one obtains

$$
U\alpha(n) = (NE_0 + NC - NzJ)\,\alpha(n) + J \sum_{n'} [\alpha(n) - \alpha(n')] \quad . \tag{16.40}
$$

One now removes the constant terms in (16.40) by defining a new quantity ε

$$
2J\varepsilon = U - (NE_0 + NC - NzJ) \quad ,
$$

$$
\varepsilon = \frac{U}{2J} + \text{const.} \quad , \tag{16.41}
$$

and gets

$$
2\varepsilon\alpha(n) = \sum_{n'} [\alpha(n) - \alpha(n')] \quad , \tag{16.42}
$$

where the constant terms in ε have been omitted. To solve (16.42) one assumes the following distributions

distribution n	↑	↑	↓	↑	↑
distribution n'_1	↑	↓	↑	↑	↑
distribution n'_2	↑	↑	↑	↓	↑
	$n-2$	$n-1$	n	$n+1$	$n+2$

In carrying out the sum over the n' one has to sum over the pairs of antiparallel spins in the new distributions n'. Equation (16.42) thus becomes

$$2\varepsilon\alpha\left(n\right) = \alpha\left(n\right) - \alpha\left(n-1\right) + \alpha\left(n\right) - \alpha\left(n+1\right) \quad . \tag{16.43}$$

Equation (16.43) is a differential equation, where the rhs represents the negative second derivative of $\alpha\left(n\right)$ written as finite differences. To find a solution one chooses the usual Ansatz

$$\alpha\left(n\right) = \exp\left(inka_0\right) \quad . \tag{16.44}$$

This function can be also seen as a Fourier representation of the spin lattice with the lattice constant a_0. One obtains the general solution

$$\varepsilon = 1 - \cos\left(ak_0\right) \quad . \tag{16.45}$$

Equation (16.45) describes the dispersion relation for spinwaves (magnons) as they were also found as solutions of the Heisenberg model (7.25). These are elementary excitations of the spin lattice. In Chap. 7 the Hamiltonian was diagonalized via a transformation employing boson operators. From the derivation above it becomes clear why these magnons are to be described as a particle with integer spin. Since a magnon consists at least of one pair of antiparallel spins moving through the lattice their total spin amounts to zero and thus magnons are bosons.

In a classical picture the excitation of a spinwave causes a precession of the spin system, where the phase difference between two lattice sites is determined by the wave vector $\mathbf{k}$. The uniform precession, i.e. $\mathbf{k} = 0$ (infinite wavelength) is known as ferromagnetic resonance. Figure 7.1 in Chap. 7 shows this classical picture.

In real space a magnon consists of a single pair of antiparallel spins moving through space. This single pair is described by a superposition of an infinite number of spinwaves with varying $\mathbf{k}$.

17. Exchange and Correlation in Metals

In the previous chapter it was shown how the exchange interaction is responsible for the formation of magnetic order in systems with localized spins. Unfortunately for itinerant electrons the subject is rather more complicated and one has to develop the theory for the limiting cases of the free electron gas and of tightly bound electrons.

17.1 Free Electron Gas

Although the free electrons never really carry ferromagnetism it is useful to study them as a model system. Interactions between free electrons are of long range (coulomb type). In metals the interactions are of shorter range due to the shielding effects of the other electrons, which will be discussed later. However, the results derived from the electron gas form the basis of the local density approximation (LDA) for exchange and correlation (for an excellent account of LDA and itinerant magnetism see [146]). Within this highly successful scheme, exchange and correlation are calculated from the electron density, rather than carrying out the full integration over all states in the system.

The total energy of the free electron gas is given by (8.9)

$$(8.9) \rightarrow E_e = \frac{3}{10} n \varepsilon_F \left[(1 + \zeta)^{\frac{5}{3}} + (1 - \zeta)^{\frac{5}{3}} \right] + \text{const.} \quad , \tag{17.1}$$

where the Fermi energy is given by (2.26)

$$\varepsilon_F = \frac{h^2}{2m} \left(\frac{3 N_{\text{tot}}}{8 \pi V} \right)^{\frac{2}{3}} \quad . \tag{17.2}$$

The exchange interaction of the free electron gas will be calculated within the framework of the Hartree–Fock model. The HF-model approximates the many particle problem by an effective single particle Schrödinger equation, where a single electron is assumed to move in the field of the other $n - 1$ electrons. This field has to be calculated self consistently (SCF methods). In the Hartree–Fock model the exchange integral is given by the sum over both spin directions

$$J = -\frac{1}{2} \left(\overset{\uparrow}{\sum_{ij}} J_{ij} + \overset{\downarrow}{\sum_{ij}} J_{ij} \right) \quad, \tag{17.3}$$

where

$$J_{ij} = \int \frac{e^2}{|\boldsymbol{r}_1 - \boldsymbol{r}_2|} \psi_i^* (\boldsymbol{r}_1) \psi_j^* (\boldsymbol{r}_2) \psi_i (\boldsymbol{r}_2) \psi_j (\boldsymbol{r}_1) \, \mathrm{d}\tau_1 \mathrm{d}\tau_2 \quad. \tag{17.4}$$

The wavefunction for a free electron is a plane wave, namely

$$\psi_i (\boldsymbol{r}) = \frac{1}{\sqrt{V}} \exp\left(\mathrm{i} \boldsymbol{k}_i \boldsymbol{r} \right) \quad. \tag{17.5}$$

Therewith J_{ij} can be written

$$J_{ij} = \frac{e^2}{V^2} \int \frac{\exp\left[-\mathrm{i}\left(\boldsymbol{k}_i - \boldsymbol{k}_j\right)\left(\boldsymbol{r}_1 - \boldsymbol{r}_2\right)\right]}{|\boldsymbol{r}_1 - \boldsymbol{r}_2|} \mathrm{d}\tau_1 \mathrm{d}\tau_2 \quad. \tag{17.6}$$

This integral can be evaluated by means of the Poisson equation, and gives

$$J_{ij} = \frac{4\pi e^2}{V} \frac{1}{|\boldsymbol{k}_i - \boldsymbol{k}_j|^2} \quad. \tag{17.7}$$

Integrating over all occupied states with wavevectors $\boldsymbol{k}_i$ and $\boldsymbol{k}_j$ yields

$$-\overset{\uparrow\downarrow}{\sum_{\boldsymbol{k}_i, \boldsymbol{k}_j}} = -\frac{3}{2} e^2 \left(\frac{3}{4\pi V} \right)^{\frac{1}{3}} \left(n^{\pm} \right)^{\frac{4}{3}} \quad, \tag{17.8}$$

with $n^{\pm} = \frac{1}{2} n \left(1 \pm \zeta \right)$, and finally

$$J = -\frac{3}{8} n \varepsilon_{\mathrm{J}} \left[(1 + \zeta)^{\frac{4}{3}} + (1 - \zeta)^{\frac{4}{3}} \right] \quad, \tag{17.9}$$

with

$$\varepsilon_{\mathrm{J}} = e^2 \left(\frac{3n}{\pi V} \right)^{\frac{1}{3}} \quad. \tag{17.10}$$

Equation (17.9) describes the exchange energy for the free electron gas with Coulomb interaction. One finds that this energy is proportional to the electron density (n/V). This result is the starting point for the LDA (local density approximation). In the LDA one calculates the ground state properties of a many particle system without the antisymmetry condition but including a correction for the exchange interaction based on the result given above. Adding the exchange term to (17.1) yields the total energy as given by

$$\frac{U}{n\varepsilon_{\mathrm{F}}} = \frac{3}{10} \left[(1 + \zeta)^{\frac{5}{3}} + (1 - \zeta)^{\frac{5}{3}} \right] - \frac{3}{8} \frac{\varepsilon_{\mathrm{J}}}{\varepsilon_{\mathrm{F}}} \left[(1 + \zeta)^{\frac{4}{3}} + (1 - \zeta)^{\frac{4}{3}} \right] \quad. \tag{17.11}$$

Figure 17.1 shows a plot of the total energy $\frac{U}{n\varepsilon_{\mathrm{F}}}$ as a function of ζ. The result is quite surprising and sobering.

For curve A , $\frac{\varepsilon_J}{\varepsilon_F} < 2$, the equilibrium state is found for $\zeta = 0$ (paramagnetism). For curve B, $\frac{\varepsilon_J}{\varepsilon_F} > 2$, and results in a fully polarized system $\zeta = 1$. This is the condition for ferromagnetic order in a system with purely coulomb type interaction. Within the Hartree–Fock approximation there exist only two possible solutions for the free electron gas. The reason is found in the fact that the exchange term is exactly of coulomb form and therefore has an infinite range of interaction whereas in real metals the exchange interaction is shielded by the other electrons so that its range of interaction is reduced. Only with this shielding can weak itinerant magnetism occur.

It should be noted that the HF approximation only accounts for exchange but not for correlation effects. Considering also correlation is a rather complicated task. Correlation means that these electrons do not move completely independently from one another. In metallic systems correlation effects should be smaller than exchange effects; nevertheless there is a rising interest in highly correlated systems (ceramic superconductors, Mott insulators) where the correlation correction affects the whole physics. As correlation occurs only between antiparallel spins it becomes zero for $\zeta = 1$. For $\zeta = 0$ the correlation energy is negative and thus reduces exchange effects.

17.2 Tightly Bound Electrons

In the above, the electrons in a metal were approximated by a free electron gas. Now we move to the other extreme and describe the electrons by atomic-like functions (Wannier functions) which are always localized on a single atomic site (tight binding approximation). The respective wavefunction reads

$$\psi_i (r_i) = \frac{1}{\sqrt{N}} \sum_l \exp (i k_i R_l) \, \phi (r_i - R_l) \quad , \tag{17.12}$$

where N is the number of electrons (or atoms), R the position of the atom with number l and $\phi (r_i - R_l)$ the atomic wavefunction. For both spins the

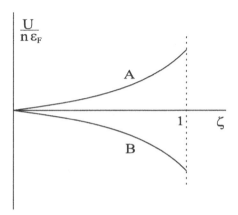

Fig. 17.1. Hartree–Fock total energy as a function of ζ

exchange integral is again given by (17.3) and (17.4). Inserting the wavefunction (17.12) one obtains

$$J_{ij} = \frac{1}{N^2} \sum_{l,m,n,p} I\left(l, m, n, p\right) \exp\left[-i\left(k_i\left(R_l - R_p\right) + k_j\left(R_m - R_n\right)\right)\right] ,$$

(17.13)

where

$$I\left(l, m, n, p\right) = \int \frac{e^2}{|r_1 - r_2|} \phi^*\left(r_1 - R_l\right) \phi^*\left(r_2 - R_m\right)$$
$$\times \phi\left(r_1 - R_n\right) \phi\left(r_2 - R_p\right) d\tau_1 d\tau_2. \quad (17.14)$$

The integral involves two electrons and four centers. These integrals are well known in quantum chemistry (but nevertheless are very complicated). A slight simplification can be obtained by putting the atom l at the origin of the coordinate system ($l = 0$) and by restricting the integration to the next-nearest-neighbors. One obtains

$$J_{ij} = \frac{1}{N^2} \sum_{m,n,p} I\left(0, m, n, p\right) \exp\left[-i\left(-k_i R_p + k_j\left(R_m - R_n\right)\right)\right]. \quad (17.15)$$

In a symbolic notation $m, n, p = 1$ for a neighbor of the atom $l = 0$, all contributions to J_{ij} are given in the table below. The following abbreviations are used:

- R: distance between the neighbors.
- $m, n, p, = R_m, R_n, R_p$ for all neighbors which are positioned symmetrically around the central atom.
- $\exp\left[-i\left(-k_i R_p + k_j\left(R_m - R_n\right)\right)\right] = \exp\left(-iA\right).$
- $\int \frac{e^2}{|r_1 - r_2|} \phi^*\left(r_1\right) \phi^*\left(r_2 - R_m\right) \phi\left(r_1 - R_n\right) \phi\left(r_2 - R_p\right) d\tau_1 d\tau_2$
$= \int \frac{e^2}{|r_1 - r_2|} B d\tau_1 d\tau_2$

m	n	p	A	B	Integral
0	0	0	0	$\left|\phi\left(r_1\right)\right|^2 \left|\phi\left(r_2\right)\right|^2$	I_0
1	0	0	$k_j R$	$\left|\phi\left(r_1\right)\right|^2 \phi\left(r_2\right) \phi^*\left(r_2 - R\right)$	I_2
0	1	0	$-k_j R$	$\left|\phi\left(r_2\right)\right|^2 \phi\left(r_1\right) \phi^*\left(r_1 - R\right)$	I_2
0	0	1	$-k_i R$	$\left|\phi\left(r_1\right)\right|^2 \phi\left(r_2\right) \phi^*\left(r_2 - R\right)$	I_2
1	1	0	0	$\phi^*\left(r_1\right) \phi\left(r_2\right) \phi^*\left(r_2 - R\right) \phi\left(r_1 - R\right)$	I_3
1	0	1	$-k_i R - k_j R$	$\phi^*\left(r_1\right) \phi^*\left(r_2\right) \phi\left(r_2 - R\right) \phi\left(r_1 - R\right)$	I_3
0	1	1	$-k_i R + k_j R$	$\left|\phi\left(r_1\right)\right|^2 \left|\phi\left(r_2 - R\right)\right|^2$	I_1
1	1	1	$-k_i R$	$\phi^*\left(r_1\right) \phi\left(r_1 - R\right) \left|\phi\left(r_2 - R\right)\right|^2$	I_2

For z nearest-neighbors on a cubic lattice one thus finds

$$J_{ij} = \frac{1}{N}\Big[I_0 + zI_1\left(R\right) \cos\left(k_i R\right) \cos\left(k_j R\right) + 4zI_2\left(R\right) \cos\left(k_i R\right)$$
$$+ zI_3\left(R\right) + zI_3\left(R\right) \cos\left(k_i R\right) \cos\left(k_j R\right)\Big]. \quad (17.16)$$

The integrals I_0, I_1, I_2, I_3 are one- and two-center integrals which are again well known (and often tabulated) in quantum chemistry. Integrals of this type occur quite frequently and, for special basis sets like Gaussians, can also be evaluated analytically. Their properties are now discussed in more detail.

1. $$I_0 = \int \frac{e^2}{|\boldsymbol{r}_1 - \boldsymbol{r}_2|} |\phi(\boldsymbol{r}_1)|^2 |\phi(\boldsymbol{r}_2)|^2 \, d\tau_1 d\tau_2 \quad .$$

    I_0 is an intraatomic Coulomb integral of two electrons on one site. For pure $3d$-functions one finds a very large value of about 20eV. In a real $3d$ metal the screening by the $4s$-electrons reduces the value to about 3eV. For degenerate electrons there exist a number of these integrals for different sets of quantum numbers. These exchange integrals are called Hund-terms and are important in the theory of atomic spectra, their magnitude is about 1eV. One finds that all these effects enter somehow into the exchange term. One usually formulates an interaction I_b (bare interaction) which due to the Hund-, screening-, and correlation-terms will be reduced to an effective interaction I_{eff}.

2. The integral I_1 is an interatomic integral between two atoms on different sites

    $$I_1 = \int \frac{e^2}{|\boldsymbol{r}_1 - \boldsymbol{r}_2|} |\phi(\boldsymbol{r}_1)|^2 |\phi(\boldsymbol{r}_2 - \boldsymbol{R})|^2 \, d\tau_1 d\tau_2 \quad .$$

    For pure $3d$-functions one finds a value of about 4eV, which is again reduced by screening effects to 1eV.

3. The integrals I_2 and I_3 are also interatomic ones, but give only very small contributions. In particular I_3 is of a form similar to one which can be derived for localized systems and should thus be small in our itinerant case.

To calculate the total energy of the lattice in the Hartree–Fock model one also has to determine the respective coulomb term which is given

$$C = \frac{1}{2} \sum_{ij} C_{ij} \quad , \tag{17.17}$$

$$C_{ij} = \frac{1}{N} \Big[I_0 + z I_1(\boldsymbol{R}) + 4z I_2(\boldsymbol{R}) \cos(\boldsymbol{k}_i \boldsymbol{R})$$
$$+ 2z I_3(\boldsymbol{R}) \cos(\boldsymbol{k}_i \boldsymbol{R}) \cos(\boldsymbol{k}_j \boldsymbol{R}) \Big] \quad . \tag{17.18}$$

Again C and J involve summation over the occupied states. One sees immediately that for $|\boldsymbol{R}| \to \infty$ both C_{ij} and J_{ij} have the same value

$$C_{ij} = J_{ij} = \frac{I_0}{N}$$

One has to consider the two cases for $\zeta = 1$ and $\zeta = 0$ for $|\boldsymbol{R}| \to \infty$

- $\zeta = 1$; ferromagnetism, all spins are parallel and hence $U = N E_0$, because C and J cancel each other.

- $\zeta = 0$; paramagnetism, C and J do not cancel each other. In calculating J one assumes that half of the spins are spin-up and half are spin-down so that the total energy becomes

$$U - NE_0 = \frac{1}{2}\left(NI_0 - 2\frac{N^2}{4}\frac{1}{N}I_0\right) = \frac{1}{4}NI_0 \quad . \tag{17.19}$$

Plotting $U - NE_0$ as a function of $|\boldsymbol{R}|$ (Fig. 17.2) for both cases shows a surprising result, namely that in Hartree–Fock theory for infinite separation of the electrons the ferromagnetic state is always more stable. This error is buried in the fact that the HF wavefunction is assumed to be a product of single particle wavefunctions. This product gives reliable results as long as the atomic distance is small so that a product of wavefunctions is reasonable. For large distances a linear combination (sum) of single particle wavefunctions would meet reality much better. This is the case for the Heitler–London model (Sect. 16.1) where for infinite separation the energy is always $U = NE_0$ (which is the sum of the energies of N non-interacting particles) independent of the mutual orientation of the spins and thus independent of the value of ζ. As the correlation energy is zero for $\zeta = 1$, the error in the Hartree–Fock model is due to the missing correlation term. From the difference between the two cases the correlation energy can be calculated to be

$$U_{\text{corr}} = -\frac{1}{4}NI_0 \quad \text{for} \quad \zeta = 0 \quad , \tag{17.20}$$

$$U_{\text{corr}} = 0 \quad \text{for} \quad \zeta = 1 \quad . \tag{17.21}$$

For finite values of $\boldsymbol{R}$ the calculation of the exchange integral becomes very complicated. The usual way to circumvent this problem is to assume that the major contribution to I_0 does not depend of $\boldsymbol{k}$. For practical application one introduces a quantity I_{eff} which also contains correlation effects.

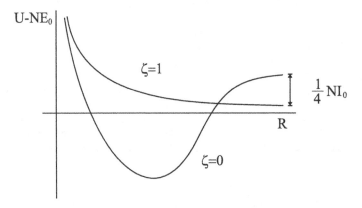

Fig. 17.2. Hartree–Fock energy for the ferromagnetic ($\zeta = 1$) and the paramagnetic ($\zeta = 0$) case

$$U_{xc} = \frac{1}{N}\left(J^+ + J^-\right)$$
$$= -\frac{1}{2}I_{eff}\left[\left(n^+\right)^2 + \left(n^-\right)^2\right] \quad , \tag{17.22}$$
$$n^{\pm} = \frac{n}{2}\left(1 \pm \zeta\right) \quad .$$

This yields an expression which is also known from the Hubbard–model

$$\frac{1}{N}\left(J^+ + J^-\right) = -\frac{1}{2}I_{eff}\, n^2 + I_{eff}\, n^+n^- \quad . \tag{17.23}$$

Using (17.22) one can also arrive at a direct comparison with the molecular field term which was postulated for the Stoner model

$$U_{xc} = -\frac{1}{2}I_{eff}\frac{1}{4}n^2\left[(1+\zeta)^2 + (1-\zeta)^2\right] \tag{17.24}$$
$$= -\frac{1}{2}nk_B\Theta\zeta^2 + \text{const.}$$

With the usual formulation that $k_B\Theta = \frac{1}{2}nI_{eff}$, (17.24) provides an interpretation of the molecular field in terms of the exchange interaction. This is also the reason why these molecular fields have such large (almost unphysically large) values. The exchange interaction is an electrostatic interaction between electrons which occurs on a completely different energy scale that usual magnetic fields. Expressing the respective Coulomb energy in terms of a magnetic field leads to extremely large values, such as those given in Table 6.1.

18. Spin Fluctuations

In Chap. 12 the temperature dependence of the magnetic properties was calculated as arising from the temperature dependence of the Fermi distribution. The results of this model were rather disappointing and led to an unrealistic description of the magnetic behavior. For localized moments the Heisenberg model (Chap. 7) and similar approaches yield fair results. For these models collective excitations of the whole spin system were formulated earlier. In the case of itinerant electrons one can no longer assume the existence of a localized spin and statistical fluctuations of the magnetic moment (the magnetization-density/spin-density) have to replace the spinwaves. Spin fluctuation theory in the way it has been introduced by Moriya [43] has added much to our understanding of itinerant ferromagnetism. The Landau–Ginzburg theory presented here (and in detail in Chap. 20) was introduced by Murata and Doniach [44] and later quite successfully applied by Lonzarich at al. [45, 46] to explain the properties of weakly itinerant ferromagnets Ni_3Al and $MnSi$.

In an itinerant electron system the magnetic moment is carried by the spin-density. Figure 18.1 sketches how such a fluctuation of the spin-density can be envisaged.

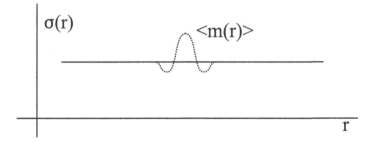

Fig. 18.1. Sketch of a fluctuation in the spin-density $\sigma(r)$

18.1 Fluctuations of a Thermodynamical Variable

Those physical quantities that characterize the equilibrium state of a macroscopic system are always very close (with high accuracy) to their average value (this fact is an alternative definition of equilibrium). However, there are – small – deviations from equilibrium, the variables fluctuate and one has the problem of finding the probability distribution for these deviations. One assumes a system in equilibrium and x to be a certain physical quantity which characterizes the system as a whole or a part of it (in the first case, x must not be a quantity which has to be constant due to any conservation law: e.g. the total energy). An elegant way to formulate fluctuations is to use Gaussian statistics [147].

From Boltzmann's formula one obtains for the probability $w(x)$

$$w(x) = \exp\left(S(x)\right) \times \text{const.} \tag{18.1}$$

Since the entropy $S(x)$ has to be a maximum for the equilibrium one finds that for $x = \langle x \rangle = 0$ ($\langle x \rangle$ is the average value of x) the first derivative of $S(x)$ must vanish and the second derivative be negative.

$$\frac{\partial S(x)}{\partial x} = 0 \quad , \quad \frac{\partial^2 S(x)}{\partial x^2} < 0 \quad \text{for } x = 0 \quad . \tag{18.2}$$

It should be noted that these assumptions of equilibrium properties for the entropy causes trouble when approaching a critical point where both the first and the second derivative vanish. The Gaussian statistics for the fluctuations are therefore restricted to those parts of phase space that are far from critical points.

Since the quantity x should be small, one expands $S(x)$ in a Taylor series up to second order, where the linear term vanishes because of the condition formulated in (18.2)

$$S(x) = S(0) - \frac{\beta}{2}x^2 \quad . \tag{18.3}$$

With (18.1) one obtains

$$w(x)\,\mathrm{d}x = \kappa \exp\left(-\frac{\beta}{2}x^2\right)\mathrm{d}x \quad . \tag{18.4}$$

The constant κ is given by the condition $\int w(x)\,\mathrm{d}x = 1$. Although the expression for $w(x)$ is only valid for small x, one can carry out the integration from $-\infty$ to $+\infty$ because the integrand vanishes rapidly for larger values of x. The normalization constant κ is thus given by

$$\kappa = \left(\frac{\beta}{2\pi}\right)^{\frac{1}{2}} \quad . \tag{18.5}$$

The probability distribution for x is now given by

$$w\left(x\right)\mathrm{d}x = \left(\frac{\beta}{2\pi}\right)^{\frac{1}{2}}\exp\left(-\frac{\beta}{2}x^2\right)\mathrm{d}x \quad . \tag{18.6}$$

A distribution of this kind is called a Gaussian distribution. It shows a maximum for $x = 0$ and decreases monotonically for positive and negative values of x. The statistical averages for even powers of the variable x are now given by

$$\left\langle x^{2k}\right\rangle = \left(\frac{\beta}{2\pi}\right)^{\frac{1}{2}}\int\limits_{-\infty}^{+\infty} x^{2k}\exp\left(-\frac{\beta}{2}x^2\right)\mathrm{d}x = \beta^{-k}\left(2k-1\right)!! \quad . \tag{18.7}$$

This means in particular that $\left\langle x^2\right\rangle = \beta^{-1}$ and $\left\langle x^4\right\rangle = 3\beta^{-2} = 3\left\langle x^2\right\rangle^2$. Each higher order average can be expressed in terms of $\left\langle x^2\right\rangle$.

18.2 Fluctuations of the Magnetic Moment

In the framework of the Landau theory of phase transitions, the magnetic moment M was introduced as the order parameter (14.10). Without further discussion it was assumed that M is a scalar. This was possible because M appeared only in even powers as $M^{2n} = \left(M \cdot M\right)^n$ which are always scalars. When one now deals with fluctuations one has to recognize the vector properties of both the magnetization and its fluctuations. For reasons of symmetry one can postulate that the volume integral over odd powers in the fluctuation always vanishes. If now $m\left(r\right)$ is the locally fluctuating magnetic moment one assumes

$$\frac{1}{V}\int\left(m\left(r\right)\right)^n\mathrm{d}\tau = \begin{cases} \left\langle m^n\right\rangle & \text{for } n = 2k \\ 0 & \text{for } n = 2k+1 \end{cases} \quad . \tag{18.8}$$

As an example the lowest two powers of the total magnetic moment are calculated. Fluctuations appear in all three space directions. The direction of $M = (0, 0, M_z)$ is taken as the z axis of a local coordinate system, so that one has to consider two fluctuations perpendicular to M and one parallel. These three components of the fluctuation are denoted $m_i = m_i e_i$ where $(m_1, m_2, m_3) = (m_x, m_y, m_z)$ are the three components in the directions given by the unit vectors e_i. (m_z is parallel to M, m_x and m_y are perpendicular to M). One now extends the original order parameter M^{2n} by the statistical average including the fluctuation terms. The bulk magnetization M remains the order parameter which of course will vanish at the critical temperature (Curie temperature).

$$M^{2n} \rightarrow \left\langle\left(M + \sum_{i=1}^{3} m_i\right)^{2n}\right\rangle \quad . \tag{18.9}$$

Putting $n = 1$ yields

$$\left\langle \left(M + \sum_{i=1}^{3} m_i \right)^2 \right\rangle = \left\langle M^2 + 2M \sum_{i=1}^{3} m_i + \sum_{i=1}^{3} \sum_{j=1}^{3} m_i m_j \right\rangle$$

$$= M^2 + 2 \left\langle m_\perp^2 \right\rangle + \left\langle m_\parallel^2 \right\rangle \quad . \tag{18.10}$$

On averaging, the mixed term vanishes, because m_i appears as an odd power. For the same reason in the term $\sum_{i=1}^{3} \sum_{j=1}^{3} m_i m_j$ for the same reason only the diagonal elements remain. On the rhs of (18.10) it was assumed that the system is isotropic. This means that one describes one component $\left\langle m_\parallel^2 \right\rangle$ which is parallel to the direction of M and two components $\left\langle m_\perp^2 \right\rangle$ which are equal and perpendicular to M.

Putting $n = 2$ yields

$$\left\langle \left(M + \sum_{i=1}^{3} m_i \right)^4 \right\rangle = \left\langle \left(M^2 + 2M \sum_{i=1}^{3} m_i + \sum_{i=1}^{3} \sum_{j=1}^{3} m_i m_j \right)^2 \right\rangle$$

$$= \left\langle \left(M^2 + 2M m_3 + \sum_{i=1}^{3} \sum_{j=1}^{3} m_i m_j \right)^2 \right\rangle$$

$$= \left[\left\langle M^4 + 4M^2 m_3^2 + \sum_{i,j,k,l=1}^{3} m_i m_j m_k m_l \right. \right.$$

$$\left. \left. + 4M^3 m_3 + 2M^2 \sum_{i=1}^{3} \sum_{j=1}^{3} m_i m_j + 4M m_3 \sum_{i=1}^{3} \sum_{j=1}^{3} m_i m_j \right\rangle \right]$$

$$= M^4 + M^2 \left(6 \left\langle m_\parallel^2 \right\rangle + 4 \left\langle m_\perp^2 \right\rangle \right) + 8 \left\langle m_\perp^2 \right\rangle^2$$

$$+ 3 \left\langle m_\parallel^2 \right\rangle^2 + 4 \left\langle m_\perp^2 \right\rangle \left\langle m_\parallel^2 \right\rangle \quad . \tag{18.11}$$

An analytical recipe for calculating the coefficients of the various polynomial terms which appear in (18.10) and (18.11) as well as for the coefficients up to order $n = 6$ is given in Sect. J. To enter the fluctuation terms into the free energy one replaces the bulk magnetization M in (14.10) by the respective averages (18.10) and (18.11). One obtains an expression for the free energy in the variable M which describes the macroscopic magnetic moment (bulk moment) and the variables $\left\langle m_\perp^2 \right\rangle$ and $\left\langle m_\parallel^2 \right\rangle$ giving the quadratic average over the locally fluctuating magnetic moment. To distinguish the free energy, which includes the fluctuation terms, from the straightforward Landau expansion one sometimes uses the name "dynamical" form of (14.10) which now reads

$$F = \frac{A}{2}\left(M^2 + 2\langle m_\perp^2 \rangle + \langle m_\parallel^2 \rangle\right)$$
$$+\frac{B}{4}\left[M^4 + M^2\left(6\langle m_\parallel^2 \rangle + 4\langle m_\perp^2 \rangle\right)\right.$$
$$\left.+8\langle m_\perp^2 \rangle^2 + 3\langle m_\parallel^2 \rangle^2 + 4\langle m_\perp^2 \rangle\langle m_\parallel^2 \rangle\right] \quad . \tag{18.12}$$

The static fluctuations-free form of the free energy has just been extended by the fluctuations. For $T = 0$ the fluctuation amplitude is zero and (18.12) reduces to (14.10). One also notices that the dynamics of the fluctuations scales with the static susceptibility (contained in the coefficient A). The first important result is for the fluctuation amplitude at T_c. In the framework of the Landau theory, the Curie temperature is given by the point where the susceptibility diverges under the condition $M = 0$. As the bulk magnetization M vanishes at T_c there is no longer a difference between the parallel and perpendicular fluctuations. This means that for $T \geq T_c \Rightarrow \langle m_\parallel^2 \rangle = \langle m_\perp^2 \rangle = \langle m^2 \rangle$. Calculating $\frac{d^2F}{dM^2} = \chi^{-1}$, with $M = 0$, one obtains

$$\chi^{-1} = A + \frac{B}{2}10\langle m^2 \rangle = 0 \quad \text{for} \quad T = T_c \quad , \tag{18.13}$$

which determines the magnitude of the fluctuations at T_c. One recovers the so called Moriya formula

$$\langle m_c^2 \rangle = \frac{M_0^2}{5} \quad \text{for} \quad T = T_c \quad , \tag{18.14}$$

where $\langle m_c^2 \rangle$ is the fluctuation amplitude at the critical temperature. Equation (18.14) shows that this amplitude is entirely given by ground state quantities defined at $T = 0$. Equation (18.14) also allows to formulate an approximation for the temperature dependence of the fluctuations. From the fluctuation dissipation theorem [147] it is obvious that classical fluctuations change essentially linear with temperature. Since the fluctuations have to vanish at $T = 0$ and their value at $T = T_c$ is known, one can approximate the temperature variation by

$$\langle m_\parallel^2 \rangle(T) = \langle m_\perp^2 \rangle(T) \simeq \langle m_c^2 \rangle \frac{T}{T_c} = \frac{M_0^2}{5}\frac{T}{T_c} \quad . \tag{18.15}$$

In (18.15) it was also assumed that $\langle m_\parallel^2 \rangle$ and $\langle m_\perp^2 \rangle$ are identical, which is largely true for isotropic systems.

From (18.12) one can directly calculate the temperature dependence of the bulk moment M

$$\frac{dF}{dM} = 0 = AM + BM^3 + BM\left(3\langle m_\parallel^2 \rangle + 2\langle m_\perp^2 \rangle\right) \quad ,$$
$$\Rightarrow M^2 = M_0^2 - 3\langle m_\parallel^2 \rangle - 2\langle m_\perp^2 \rangle \quad . \tag{18.16}$$

Using the approximations given in (18.15) one thus obtains

$$M^2 = M_0^2 \left(1 - \frac{T}{T_c}\right) \quad . \tag{18.17}$$

This behavior differs from the Stoner model insofar as the reduction of the magnetic moment at low temperature is stronger. The reason is the same as for the comparison between the Weiss and the Heisenberg models: the collective modes described by the spin fluctuations can readily be excited at low temperature, where the Stoner excitations are still very small. A comparison of these two models is shown in Fig. 18.2.

It is also interesting to study the temperature dependence of the total magnetic moment defined in (18.10)

$$\left\langle \left(M + \sum_{i=1}^{3} m_i \right)^2 \right\rangle = M_0^2 \left(1 - \frac{T}{T_c}\right) + 2\langle m_\perp^2 \rangle + \langle m_\parallel^2 \rangle$$

$$\simeq \begin{cases} M_0^2 \left(1 - \frac{2}{5}\frac{T}{T_c}\right) & T < T_c \\ M_0^2 \frac{3}{5}\frac{T}{T_c} & T \geq T_c \end{cases} \quad . \tag{18.18}$$

The temperature behavior of the bulk moment, the fluctuations and the order parameter is shown in Fig. 18.3.

The susceptibility below and above T_c is given by

$$\chi = \chi_0 \left(1 - \frac{T}{T_c}\right)^{-1} \qquad \text{for } T < T_c \quad , \tag{18.19}$$

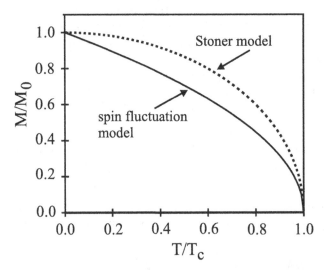

Fig. 18.2. Temperature dependence of the magnetization for the spin-fluctuation model (full curve) and the Stoner model (dashed curve)

$$\chi = 2\chi_0 \left(\frac{T}{T_c} - 1\right)^{-1} \quad \text{for } T \geq T_c \quad , \qquad (18.20)$$

which leads to a Curie constant C

$$C = \frac{d\chi^{-1}}{dT} = \frac{1}{2\chi_0 T_c} \quad , \qquad (18.21)$$

which is no longer temperature dependent (as for the Stoner model, see 12.41)) and thus describes a Curie–Weiss behavior. This result can be used to derive an expression for the effective magnetic moment 6.14 created by the spin fluctuations. Equating the definition of the Curie constant in terms of the effective magnetic moment 6.16 with 18.21 yields

$$\mu_{\text{eff}}^2 = \frac{3k_B}{2\chi_0 T_c} \quad . \qquad (18.22)$$

This allows to explain the T_c-dependence found for the Rhodes–Wohlfarth plot (Fig. 6.5). By identifying the effective carriers as being proportional to q_c one immeadiatley recovers that

$$\frac{q_c}{q_0} \propto \frac{1}{\sqrt{T_c}} \quad , \qquad (18.23)$$

which nicely describes the T_c-dependence of this otherwise phenomenologically derived relation.

Based on these results and the work by Lonzarich and Taillefer [46], Mohn and Wohlfarth [47] derived a simple model for the Curie temperature in weakly itinerant systems

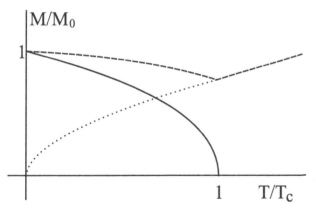

Fig. 18.3. Temperature dependence of the bulk moment $\sqrt{\left(1 - \frac{T}{T_c}\right)}$ (full curve), the fluctuations $\sqrt{\frac{3}{5}\frac{T}{T_c}}$ (dotted curve), and the total moment $\sqrt{\left(1 - \frac{2}{5}\frac{T}{T_c}\right)}$ (for $T \leq T_c$) and $\sqrt{\frac{3}{5}\frac{T}{T_c}}$ (for $T > T_c$) (dashed curve)

$$T_c \propto T_{SF} = \frac{M_0^2}{10k_B\chi_0} \quad .$$

(18.24)

In (18.24) T_{SF} is a characteristic temperature which scales with the actual ordering temperature T_c. In Fig. 18.4 the relation between T_{SF} and the experimental Curie temperature T_c is shown. The plot shows that in general T_{SF} scales rather well with T_c. From the deviations, however, one can draw conclusion about the more localized or itinerant behavior of the systems shown. Roughly speaking one finds the localized moment systems below the dashed line and the itinerant ones above.

Equation (18.24) describes the long wave length limit for fluctuations in weakly itinerant systems and is thus an approximation which is based on the assumption of an effective cut–off for the collective excitations which should be fairly constant. In the literature this approach is known as the Mohn–Wohlfarth model and has considerable interest in the application of spin fluctuation theory. Although this approach can be expected to break down for strong ferromagnets [148] it can still be used to describe the relative change of T_c, e.g. for different concentrations in an alloy system [149].

From (18.24) it also becomes clear why the Curie temperature of Co is larger, although the magnetic moment is smaller, than that for Fe. The reason is that Fe is just on the verge of a strongly ferromagnetic system, but still has a relatively large susceptibility. Co however is already strongly ferromagnetic and the susceptibility is drastically reduced. This means that in (18.24) the term $\frac{M_0^2}{\chi_0}$ (and thus T_{SF}) is smaller for Fe than for Co. For Ni the magnetic moment drops to $0.6\mu_B$, so that for a comparable susceptibility with Co the Curie temperature is reduced as well.

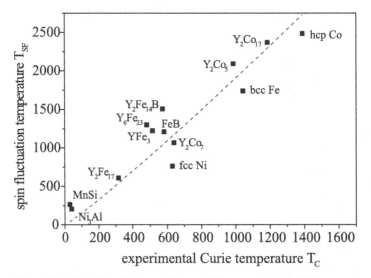

Fig. 18.4. Relation between T_{SF} and the experimental T_c

18.3 Specific Heat of the Spin Fluctuations

The expression for the free energy (18.12) allows one to calculate the specific heat using the thermodynamic relation

$$c_m = -T\frac{d^2 F}{dT^2} \quad .$$

(18.25)

If one neglects the temperature dependence of the coefficient A (neglecting single particle excitations) one obtains for $T < T_c$ the result

$$c_m^< = \frac{B}{2}\left\langle m_\parallel^2\right\rangle\left[6\frac{\partial\left\langle m_\parallel^2\right\rangle}{\partial T}+4\frac{\partial\left\langle m_\perp^2\right\rangle}{\partial T}\right]+B\left\langle m_\perp^2\right\rangle\left[2\frac{\partial\left\langle m_\parallel^2\right\rangle}{\partial T}-2\frac{\partial\left\langle m_\perp^2\right\rangle}{\partial T}\right] .$$

(18.26)

For $T \geq T_c$, the bulk magnetization M and its derivatives are zero so that one finds

$$c_m^\geq = -\frac{B}{2}\left\langle m^2\right\rangle 15\frac{\partial\left\langle m^2\right\rangle}{\partial T} \quad .$$

(18.27)

This negative contribution to the total specific heat has already been noticed by Murata and Doniach [44]. An experimental verification of this peculiar behavior was found for the Invar system Fe-Pt [150]. The specific heat of the fluctuations thus shows a discontinuity at T_c which is given by

$$\triangle c_m = \frac{B}{2}\left\langle m_c^2\right\rangle 15\frac{\partial\left\langle m^2\right\rangle}{\partial T}\bigg|_{T=T_c} \quad .$$

(18.28)

Using (18.14) and (18.15) one obtains

$$\triangle c_m = \frac{M_0^2}{4\chi_0 T_c} \quad .$$

(18.29)

For bcc Fe one calculates a value of about 36 J/mole K which is close to the experimental value of 43 J/mole K. Equation (18.29) should be compared to the analogous expression derived for the Stoner theory (14.18). One finds that the discontinuity at T_c is reduced by a factor 4 (!) with respect to the Stoner theory. The reason for this is that in the spin fluctuation model the state above T_c is not the *non-magnetic* state (like in the Stoner model) but a *paramagnetic* state, where local moments still exist, but long range order has been destroyed by the fluctuations [151].

18.4 Magneto–Volume Coupling

In analogy with Chap.14 (14.21) the interaction between the volume of a system and its magnetic moment is taken into account via a magneto–volume coupling constant δ. With these extensions the free energy reads

$$F = \frac{A}{2}\left(M^2 + 2\left\langle m_\perp^2\right\rangle + \left\langle m_\parallel^2\right\rangle\right)$$

$$+\frac{B}{4}\left[M^4 + M^2\left(6\left\langle m_\parallel^2\right\rangle + 4\left\langle m_\perp^2\right\rangle\right)\right. \tag{18.30}$$

$$\left.+8\left\langle m_\perp^2\right\rangle^2 + 3\left\langle m_\parallel^2\right\rangle^2 + 4\left\langle m_\perp^2\right\rangle\left\langle m_\parallel^2\right\rangle\right]$$

$$+\alpha + \beta V + \gamma V^2 + \delta V\left(M^2 + 2\left\langle m_\perp^2\right\rangle + \left\langle m_\parallel^2\right\rangle\right) \quad . \tag{18.31}$$

The magnetic and mechanical equations of state are given by

$$H = \left.\frac{\partial F}{\partial M}\right|_V \tag{18.32}$$

$$= AM + BM^3 + BM\left(3\left\langle m_\parallel^2\right\rangle + 2\left\langle m_\perp^2\right\rangle\right) + 2\delta V M \quad ,$$

$$-P = \left.\frac{\partial F}{\partial V}\right|_M \tag{18.33}$$

$$= \beta + 2\gamma V + \delta\left(M^2 + 2\left\langle m_\perp^2\right\rangle + \left\langle m_\parallel^2\right\rangle\right) \quad .$$

The equilibrium magnetic moment at $T = 0$ is thus

$$M_0^2 = -\frac{A + 2\delta V_0}{B} \quad , \tag{18.34}$$

where V_0 is the equilibrium volume at $T = 0$. From these results one can determine the critical pressure for the disappearance of magnetism

$$P_c = \frac{\gamma A - \beta\delta}{\delta} \quad , \tag{18.35}$$

and the pressure dependence of the Curie temperature

$$\frac{dT_c}{dP} = -\frac{T_c\left(P = 0\right)}{P_c} \quad ,$$

$$\Rightarrow T_c\left(P\right) = T_c\left(P = 0\right)\left(1 - \frac{P}{P_c}\right) \quad . \tag{18.36}$$

For fluctuation systems the Curie temperature depends linearly on the applied pressure. A discussion of the various models and their application to experiment is given in [152]. This result for the spin fluctuation model is again in contrast to Stoner theory (14.25) which gives

$$(14.25) \quad \rightarrow \quad T_c\left(P\right) = T_c\left(P = 0\right)\left(1 - \frac{P}{P_c}\right)^{\frac{1}{2}} \quad .$$

A crucial quantity is the magnetic contribution to the thermal expansion α_m since the respective value for the Stoner model was again too large. For the spin fluctuation model one finds

$$\alpha_m = \frac{d\omega}{dT} = \begin{cases} \frac{1}{T_c}\left(\frac{V_{NM}}{V_0} - 1\right)\left(1 - 3\frac{\langle m_c^2 \rangle}{M_0^2}\right) & \text{for } T < T_c \ , \\ \frac{1}{T_c}\left(1 - \frac{V_{NM}}{V_0}\right)\left(3\frac{\langle m_c^2 \rangle}{M_0^2}\right) & \text{for } T \geq T_c \ . \end{cases} \tag{18.37}$$

here ω is the relative volume change ($\frac{V-V_0}{V}$), V_0 is the equilibrium volume at $T = 0K$ and V_{NM} is the hypothetical volume of the non-magnetic phase ($M = 0, \langle m^2 \rangle = 0$), which is also the volume for the Stoner model at T_c. For the fluctuation model the volume at T_c becomes

$$V(T_c) = V_{NM} - 3(V_{NM} - V_0)\frac{\langle m_c^2 \rangle}{M_0^2} \ , \tag{18.38}$$

which leads to a considerably smaller spontaneous volume magnetostriction as given by the Stoner model, again comparable to experiment.

Equation (18.37) also allows one to recover a "rule of thumb" given by Masumoto in 1927 for Invar alloys:

• To observe a large negative value of α_m (and consequently a small overall thermal expansion) the fraction M_0^2/T_c should be large.

If one rewrites (18.37) in a different form one finds exactly Masumoto's rule

$$\alpha_m = \frac{1}{T_c}\frac{4B\delta\langle m_c^2 \rangle}{A\delta - 2B\beta} \quad \text{with } \langle m_c^2 \rangle \simeq \frac{M_0^2}{5} :$$

$$\Rightarrow \quad \alpha_m = \frac{M_0^2}{T_c}\frac{4B\delta}{5(A\delta - 2B\beta)} \tag{18.39}$$

namely that α_m is proportional to $\frac{M_0^2}{T_c}$.

18.5 Applications of the Spin Fluctuation Model

During recent years spin fluctuation theory has been successfully applied to various problems of solid state magnetism. In this section some particularly interesting ones will be briefly discussed: i) the Invar effect, ii) the temperature dependence of the susceptibility and the critical field of metamagnets, and iii) the Curie temperature of transition metal impurities in Pd.
i) Under the name Invar effect one collects various phenomena with the most striking one (which also gives the effect its name) being the vanishing thermal expansion in a certain temperature range. Among the quite large number of alloys the most prominent one is $Fe_{65}Ni_{35}$ which was already found by Guillaume [59] in the 19th century (see Fig. 5.6). Combined with this property a number of other effects appear which are more or less closely related to it. During the last 100 years there has appeared an enormous literature on the Invar effect and recently the problem was reviewed by Wassermann [153]. The basic idea how to explain the very low thermal expansion goes back to

R. Weiss [154] who proposed the so called 2-γ-state model. He assumed that in Invar alloys there exists a ferromagnetic groundstate with a larger volume and an metastable anti-ferromagnetic state with a smaller volume. As temperature rises the systems should become excited from the ferromagnetic to the anti-ferromagnetic state. As a consequence the volume would shrink. This magnetically induced shrinking of the volume should then compensate the ever-present phonon-driven thermal volume expansion. In a certain temperature range below the Curie temperature these two contributions cancel each other so that the overall thermal expansion vanishes. This model explains the basic mechanism, although the proposed anti-ferromagnetic state has never been found experimentally. Band calculations for Fe_3Ni, however, showed a very small energy difference between the magnetic and the non-magnetic state [155]. Stoner theory again failed since it predicts a magnetic contribution to the thermal expansion which is by far too negative (it would actually describe a shrinking of the crystal above T_c!). For Fe_3Ni, whose composition is close to the actual Invar alloy $Fe_{65}Ni_{35}$, it was shown how spin fluctuations lead to a smaller value for the spontaneous magnetostriction which also brings the magnetic contribution to the thermal expansion α_m to a realistic value [156]. In a second paper on this subject [157] the properties of the Fe-Ni alloy system were investigated over the full concentration range and it could be shown that α_m goes through a minimum close to the experimental concentration. In addition to the obvious success of the spin fluctuation model it was demonstrated how band structure calculations of the quantum mechanical ground state can be combined with finite temperature concepts of statistical thermodynamics to explain a rather complex phenomenon.

ii) Spin fluctuations are not restricted to ferromagnetic systems. It was also shown how the thermal properties, in particular the susceptibility, are influenced by them. For ordinary paramagnets spin-fluctuation theory leads to a Curie-Weiss behavior as expected for "localized" entities. More interesting is the case of Pd and YCo_2 where the susceptibility goes through a maximum before entering the Curie–Weiss regime. When proposing the effect of metamagnetism, Wohlfarth and Rhodes [91] considered the susceptibility maximum as a prerequisite for the possibility of such a phase transition. Again combining band structure calculations and spin fluctuation theory it could be shown that for Pd only a deviation from the linear temperature dependence of $\chi^{-1}(T)$ exists, but no discontinuous phase transition to a magnetic state. For YCo_2 there is indeed a metamagnetic phase transition at low temperatures. Above $T = 35K$ this transition is suppressed by the spin-fluctuations and the remaining phase transition is continuous [158, 159] in agreement with experiment.

iii) The finite temperature behavior of transition metal impurities in palladium [58] has also been studied within the spin fluctuation model. It was found that only small amounts of a transition metal turn the Pd-matrix to behave like a ferromagnet. Due to the very large susceptibility of the Pd-host, the transition metal impurities, which themselves form localized magnetic moments, polarize the surrounding Pd-atoms causing a polarization cloud which can extend out over more than 1000 atoms [160]. When the total moment produced that way was attributed to the impurity alone it was thought that these impurities carry "giant" moments of up to $10\mu_{\mathrm{B}}$. Experimentally the Curie temperature is not a linear function of the concentration but shows a slowing down of its increase for an impurity concentration around $> 5\%$, however, all earlier models for the Curie temperature predicted a linear concentration dependence. To explain the magnetic behavior of these systems a combined model was proposed, where the localized impurity moment is treated within a Weiss-model (see also Chap. 15). The polarizing field produced by the impurity then acts on the Pd-host which is treated as an enhanced Pauli–paramagnet with strong spin fluctuations [161]. In this sense this model describes the interaction of localized spins via a molecular field which has an additional temperature dependence from the spin fluctuations. The coupling constant between the impurity moments and the itinerant host moments can also be calculated from first principles and is in good agreement with experimental estimates. For the Curie temperature the following equation was derived

$$A + 5B\alpha k_{\mathrm{B}}T_{\mathrm{c}} + 35C\left(\alpha k_{\mathrm{B}}T_{\mathrm{c}}\right)^2 = \frac{\left(g_J - 1\right)^2}{\mu_B^2 N}K^2\frac{2J\left(J+1\right)}{3k_{\mathrm{B}}T_{\mathrm{c}}} \tag{18.40}$$

where T_{c} is given by the real solutions of this cubic equation. Equation (18.40) is identical to the result derived in (15.15) with the exception that the lhs of (18.40) represents the spin fluctuation part of the itinerant Pd-host given in terms of the spin fluctuation dependent susceptibility. The rhs describes the localized impurity moments which are given by their angular momentum J. The coupling between the localized impurity moments and the itinerant host magnetization is monitored by the coupling constant K (see (15.16) in Chap. 15). For a more detailed description of the parameters occurring in (18.40) it should be referred to the original paper. In this model only one experimental parameter comes in, namely the temperature of the susceptibility maximum for pure Pd. All other quantities which enter (18.40) were derived from first principles electronic band structure calculations.

Spin fluctuations are of course not restricted to ferromagnetic systems. Also anti-ferromagnets and ferrimagnets (which are otherwise not discussed in this book) have been treated successfully. Among these systems are γ-Mn and FeRh. FeRh is especially interesting, because with rising temperature it shows a phase transition from an anti-ferromagnetic to a ferromagnetic state. Comparing the respective free energies of these two phases allowed to determine this magnetic phase transition [213].

18.6 Comparing the Spin-Fluctuation and the Stoner-Model

In this section a comparison between the results from the spin fluctuation model and the Stoner model is presented (Table 18.1). In both cases the results are based on a Landau expansion of the free energy up to fourth order in M and up to second order in V. The starting equations are thus (14.10) or (14.21) for the Stoner model and (18.12) or (18.31) for the spin fluctuation model, respectively.

Spin fluctuations lead to a different power law behavior for both the susceptibility and the magnetic moment. However, since spin fluctuations are also treated within mean field theory, the critical exponents of both models are the same. This different power law is also found for the pressure dependence of the Curie temperature and its slope. As expected for localized moments, spin fluctuations lead to a linear temperature dependence of the inverse susceptibility (Curie–Weiss law), which is expressed by a temperature independent Curie constant. The magnetic contributions to the thermal expansion coefficient caused by the spin fluctuations are smaller than that for

Table 18.1. Comparison between the spin fluctuation and the Stoner model for the susceptibility χ, the Curie constant C, the magnetic moment $M(T)$, the magnetic contribution α_m to the thermal expansion coefficient, the volume at the Curie temperature $V(T_c)$, the pressure dependence of the Curie temperature $T_c(P)$, the coefficient dT_c/dP of the pressure dependence of T_c; the discontinuity of α_m at the Curie temperature $\Delta\alpha_m$, the discontinuity of the specific heat c_m at the Curie temperature Δc_m

	spin fluctuation model	Stoner model
χ for $T < T_c$	$\chi_0\left(1-\frac{T}{T_c}\right)^{-1}$	$\chi_0\left(1-\frac{T^2}{T_c^2}\right)^{-1}$
χ for $T \geq T_c$	$2\chi_0\left(\frac{T}{T_c}-1\right)^{-1}$	$\chi_0\left(\frac{T^2}{T_c^2}-1\right)^{-1}$
$C = \frac{\partial(\chi^{-1})}{\partial T}$	$\frac{1}{2\chi_0 T_c}$	$\frac{1}{\chi_0 T_c}\frac{T}{T_c}$
$M(T)$	$M_0\left(1-\frac{T}{T_c}\right)^{\frac12}$	$M_0\left(1-\frac{T^2}{T_c^2}\right)^{\frac12}$
α_m for $T < T_c$	$\frac{1}{T_c}\left(\frac{V_{NM}}{V_0}-1\right)\frac25$	$\frac{2}{T_c}\left(\frac{V_{NM}}{V_0}-1\right)\frac{T}{T_c}$
α_m for $T \geq T_c$	$\frac{1}{T_c}\left(1-\frac{V_{NM}}{V_0}\right)\frac35$	0
$V(T_c)$	$V_{NM}-(V_{NM}-V_0)\frac35$	V_{NM}
$T_c(P)$	$T_c(0)\left(1-\frac{P}{P_c}\right)$	$T_c(0)\left(1-\frac{P}{P_c}\right)^{\frac12}$
$\frac{\partial T_c}{\partial P}$	$-\frac{T_c}{P_c}$	$-\frac{T_c}{2P_c}$
$\Delta\alpha_m$ at T_c	$\frac{1}{T_c}\left(1-\frac{V_{NM}}{V_0}\right)$	$\frac{2}{T_c}\left(1-\frac{V_{NM}}{V_0}\right)$
Δc_m at T_c	$\frac{M_0^2}{4\chi_0 T_c}$	$\frac{M_0^2}{\chi_0 T_c}$

the Stoner model and exist also above T_c. Consequently also the volume at T_c i s larger than for the Stoner model, which causes a smaller spontaneous volume magnetostriction. Finally the discontinuity in the specific heat at T_c is also reduced by a factor 4 with respect to the Stoner model.

19. Single Particle Excitations Versus Spin Waves

This chapter tackles the question of the response of a system of itinerant electron spins to a magnetic field varying in space or time. For a homogeneous field this response function is given by the uniform susceptibility which can also be seen as generalized susceptibility $\chi(q, \omega)$ in the limit $q = 0$ (no spatial variation = infinite wavelength of the variation) and $\omega = 0$ (no time dependence). The knowledge of the dynamical properties of an electron system is of great importance not only for the understanding of its finite temperature properties, as was demonstrated for the Heisenberg model (7.1) and (7.25) where the quadratic dispersion relation found led to the Bloch $T^{\frac{3}{2}}$ law, but also for more complex magnetic order than just ferromagnetism (antiferro-, ferri-, heli-magnetism). The latter point can be understood by considering that e.g. antiferromagnetism can be seen as a magnetic order produced by a molecular field, with opposite sign at any sublattice (staggered magnetic field). Since a transition into a magnetically ordered state is due to a pole in the susceptibility (the Stoner criterion is just the condition for a pole of the uniform, $q = 0$, susceptibility), a pole of $\chi(q, \omega)$ for $q = \frac{\pi}{a_0}(0, 0, \frac{1}{2})$ would describe antiferromagnetic order along the z-axis of a cubic crystal. This simple example also demonstrates why there is no obvious "Stoner criterion" for antiferromagnetism since this would require the knowledge of the dynamical properties of the spin system, a feature which can not easily be derived from the non–magnetic ground state. The dynamical (q-dependent) susceptibility is thus a central quantity for the understanding of magnetism.

The basic excitation mechanism for an itinerant electron system is shown in Fig. 19.1. Since the single particle excitations should reduce the magnetization one assumes transitions from the majority band (spin-up) to the minority band (spin-down).

The simplified band structure given in Fig. 19.1 resembles a strong ferromagnet where the majority band is completely filled. The Fermi energy thus lies above the majority band separated from it by the Stoner gap ΔS. The spin splitting between spin-up and spin-down is given by ΔE and, for simplicity, a rigid shift is assumed. One now considers an excitation from an occupied state $k_\uparrow$ to an unoccupied state $k'_\uparrow$ where $k'_\uparrow$ can be written as $k_\downarrow + q$. If q is zero the necessary energy is exactly ΔE which simply describes ferromagnetic resonance of an electron in the respective "molecular" field. For non-zero q the response of the system is given by the dynamical susceptibility which reads [162]

$$\chi(q,\omega) = \frac{2\mu_B^2 \Gamma(q,\omega)}{1 - 2I\Gamma(q,\omega)} \quad , \tag{19.1}$$

$$\Gamma(q,\omega) = \sum_{|k|=\text{const.}} \frac{n_{k\uparrow} - n_{k'\downarrow}}{\varepsilon_{k'\downarrow} - \varepsilon_{k\uparrow} - \hbar\omega + \Delta E} \quad . \tag{19.2}$$

Equation (19.1) is of the same kind as the enhanced static susceptibility (8.21). However the density of states of the Bloch-electrons is now replaced by a generalized Lindhard-type [163] expression which for $q = 0$, $\omega = 0$ reduces to the original density of states. Formally the rhs of $\Gamma(q,\omega)$ represents a δ-function for a particular value $q \to 0$ where the density of states is then calculated by integration over a surface with $|k| = \text{const.}$, with the denominator describing the energy conservation for the creation of a single particle excitation with the energy $\hbar\omega$.

Assuming quasi-free (non-interacting) electrons in a magnetic field (the molecular field) which produces the band splitting, one can determine the major part of the excitation spectrum. For a free electron the single particle energy is given by $\frac{\hbar^2 k^2}{2m}$ so that one finds:

$$\varepsilon_k = \frac{\hbar^2 k^2}{2m} \quad ,$$

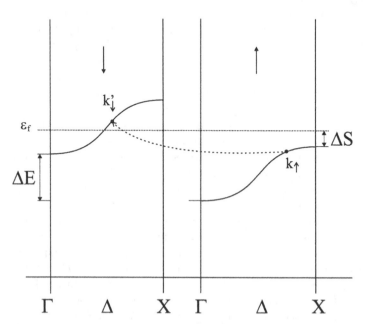

Fig. 19.1. Transfer of an electron (spin) from the majority to the minority band leading to single particle excitations in an itinerant electron system

$$\varepsilon_{k+q} = \frac{\hbar^2 (k+q)^2}{2m} + \Delta E \quad ,$$

$$\Rightarrow \hbar\omega = \frac{\hbar^2}{m}kq + \frac{\hbar^2}{2m}q^2 + \Delta E \quad . \tag{19.3}$$

For $q = 0$ one finds ferromagnetic resonance for a field given by the exchange energy $\Delta E = 2\mu_B H_M$; only for finite q does one describe single particle excitations within the Stoner continuum which is determined by all possible values of k. For the strongly ferromagnetic case considered in Fig. 19.2 these values are given by the k-range for the unoccupied states in the minority band. It is also easy to see that the smallest possible excitation energy is given by the Stoner gap ΔS. Figure 19.2 also shows how the collective excitations can readily be excited at small q values. When the collective excitations (spin waves) enter the range of the single particle excitations, they start to decay into single particle excitations (Landau damping), which reduces their lifetime drastically. As a consequence inelastic neutron scattering measures a dramatic broadening of the collective excitations (Fe: [164], Ni: [165]).

One notices immediately that the excitation of long wavelength single particle excitations always requires an energy of the order of the spin splitting since they have to reduce the z-component M_z of the magnetization.

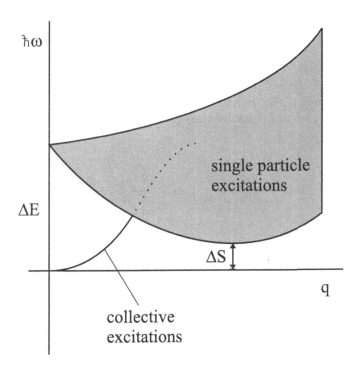

Fig. 19.2. Energy range of the single particle excitations (shaded area) and the collective excitations

Quantum fluctuations, which could produce M_x and M_y components (as for the Heisenberg model) and thus describe a tilting of the local moment are not possible in the single particle picture. However, for a weak ferromagnet, where the Stoner gap is zero so that the Stoner continuum intersects with the q-axis these single particle excitations lead to a reduction of the magnetic moment as given by (12.27). It should be noted that exactly the same single particle excitations, but only within the same spin band, describe the specific heat of an electron gas. In the strongly ferromagnetic case plotted in Fig. 19.2 the smallest possible energy for the single particle excitation is determined by the Stoner gap ΔS which gives rise to a different temperature dependence of the magnetization in the low temperature region [166]

$$\frac{M}{M_0} = 1 - T^{\frac{3}{2}} e^{-\frac{\Delta S}{k_B T}} \quad . \tag{19.4}$$

Going beyond the non-interacting electron gas and treating the effects of exchange and correlation by e.g. a Hartree–Fock Ansatz or a Hubbard model one finds an additional solution in the excitation spectrum, namely collective excitations which make up the "spinwave" branch starting from $q = 0$. These collective excitations are basic properties of the Fermi-liquid [167] and since more than one particle is involved it requires more than the single particle approximation to account for this solution. If the energy of these new states falls outside the continuum of unperturbed states (Stoner continuum) the electron–hole pairs form bound states whose energy is a smooth function of k [168].

In the case of weakly itinerant systems the energy dispersion can be calculated by expanding the real part of the generalized density of states (19.2) $\Gamma(q, \omega)$ for small q and ω. To describe poles in the dynamical susceptibility one sets the result equal to $I/2$ and obtains [169]

$$Dq^2 \simeq \hbar\omega(q) = \frac{1}{N_\uparrow - N_\downarrow}$$

$$\times \sum_k \left[\frac{1}{2} \left(n_{k_\uparrow} + n_{k_\downarrow} \right) \left(q \frac{\partial}{\partial k} \right)^2 \varepsilon_k - \frac{N \left(n_{k_\uparrow} - n_{k_\downarrow} \right)}{I \left(N_\uparrow - N_\downarrow \right)} \left(q \frac{\partial \varepsilon_k}{\partial k} \right)^2 \right] .$$

$$\tag{19.5}$$

This relation has the same quadratic dependence upon q that was found for the spin wave spectrum of a magnon in the Heisenberg model. The first term of the rhs of (19.2) corresponds to the additional kinetic energy that the electrons develop as they change their spin directions to keep up with the change in the macroscopic spin direction of the spin wave. It can be shown [170] that this effect is always positive and vanishes for a filled band. The second term corresponds to a reduction in this kinetic energy due to a tilting of the electron spin out of the plane in which the macroscopic magnetization varies. The fact that the resulting small energy change is the difference of two fairly large numbers makes it numerically rather unfavorable to evaluate this formula. One needs a highly precise description of the Fermi surface and it also requires thousands of k-points in the Brillouin-zone to obtain accurate results. However, realistic calculations involving five d-bands (rather than earlier investigations which were based on an electron gas model) have been performed for Fe [171] and Ni [172] which are in excellent agreement with experiment (Fe: [164], Ni: [165]). Applying (19.5) Edwards and Muniz [173] showed how the spinwave stiffness constant can be evaluated directly from a multi-band tight-binding parametrization of a band calculation for the ferromagnetic ground state of Fe and Ni. In the limit of the free electron gas, (19.5) can be evaluated analytically and yields

$$D = \frac{\hbar^2}{2m^*} \frac{1}{\zeta} \left[1 - \frac{2}{5} \frac{(1+\zeta)^{\frac{5}{3}} - (1-\zeta)^{\frac{5}{3}}}{(1+\zeta)^{\frac{2}{3}} - (1-\zeta)^{\frac{2}{3}}} \right] \quad , \tag{19.6}$$

which in the strongly itinerant limit $\zeta = 1$ reduces to

$$D = \frac{\hbar^2}{2m^*} \frac{4}{5} \quad , \tag{19.7}$$

where m^* is the effective mass of the electrons (see Chap. 4).

Experimentally these collective excitations have been measured already rather early by neutron diffraction [174] and show a very strong dispersion related to a large value of the spinwave stiffness constant D (7.28). The measured room temperature values for Fe, Co, and Ni are 280, 510, and 455 meVÅ^2, respectively. However, neutron experiments have the problem that for larger q-vectors these collective excitations become rapidly damped, since they decay into single particle excitations (Landau damping) once the spinwave branch enters the Stoner continuum. A different method for determining D is by fitting low temperature magnetization measurements to a $T^{3/2}$ law expected for collective excitations [175] a procedure which confirms the neutron results.

From this discussion it becomes clear that the thermal properties of itinerant electron systems should also be governed by collective excitations. Only at very high temperatures can single particle excitations become important, an effect which should be more pronounced for weak ferromagnets than for strong ones. Considering single particle excitations only, Gunnarson [176] demonstrated that the Curie temperatures obtained range between 4000 and 8000 K for Fe, Co, and Ni so that collective excitations are inevitably needed to explain the experimental values.

20. Landau–Ginzburg Model for Spin Fluctuations

In Chap. 18 spin fluctuations were introduced in a rather phenomenological way. This meant that the actual nature of these fluctuations remained unclear and that properties like the temperature dependence were only introduced in an approximative manner. Although this was sufficient to describe quite a large number of properties, here an "exact" (within the mean field approximation) theory of the magnetic fluctuations is presented. This means in particular that one has to allow for finite values of the wave-vector q. If one wants to solve the problem exactly one can follow the method of Landau and Ginzburg. The expansion for the free energy now contains a local term which takes the spatial dependence of the fluctuations into account (see also Sect. F.) and the respective Hamiltonian becomes [191]

$$\mathcal{H} = \frac{1}{V} \int d\mathbf{r} \left(E \left(M + \sum_{i=1}^{3} m_i(\mathbf{r}) \right) + \frac{C}{2} \sum_{i,j} (\nabla_j m_i(\mathbf{r}))^2 \right) \quad . \quad (20.1)$$

The Hamiltonian given by (20.1) is identical to the Ornstein–Zernicke extension (F.1) which is described in detail in Sect. F.. The term $(\nabla_j m_i(\mathbf{r}))^2$ accounts for local fluctuations. The fact that this term contains the spatial derivative of the fluctuating moment makes sure that only fluctuations which are larger that the electronic range of interaction are considered. Fluctuations on a small spatial scale are suppressed because they lead to large values of this derivative and thus to large changes in the free energy. It should be noted that the treatment of the fluctuations via a gradient term is not the only possible formulation. A highly interesting conjecture was introduced by Kirchner et al. [177] who directly introduced a non-local quadratic exchange term similar to (20.4) which couples fluctuations localized on different points in space. In a similar way Uhl et al. formulated the "exchange coupled" spin fluctuation theory [206] by assuming a mode coupling of Heisenberg type. All different approaches have in common that they assume that the long range fluctuations are responsible for the breakdown of magnetic order at finite temperature. Short ranged intraatomic fluctuations are initially suppressed and should only play any role at high temperatures. A thorough discussion of these problems can be found in [196].

$E\left(M + \sum_{i=1}^{3} m_i\left(r\right)\right)$ is the usual energy expansion as given by (18.12).
The free energy then reads

$$F = -k_{\mathrm{B}}T\ln Z \quad , \quad Z = \int \mathrm{d}\Gamma \exp\left(-\beta\mathcal{H}\right) \quad , \tag{20.2}$$

where $\mathrm{d}\Gamma$ means the integration over all fluctuation variables [see (20.7)] and
$\beta = \frac{1}{k_{\mathrm{B}}T}$. The problem is now that the sum of states cannot be calculated
analytically for the assumed Hamiltonian. A possible solution is provided by
the Bogoliubov (Peierls–Feynman) inequality which reads

$$F \le F_0 + \langle \mathcal{H} - \mathcal{H}_0 \rangle_0 \quad . \tag{20.3}$$

For a proof of (20.3) see Sect. G.. One formulates a parametrized Ansatz
for F_0 (for which the sum of states can be calculated) and finally minimizes
the rhs of (20.3). In the sense of the variational principle this leads to the
best approximation for F under the trial free energy F_0. In addition this
method leads to an effective mean field solution of the problem as discussed
in Sect. G.. To have a most general form for $\mathcal{H}_0$ one chooses a translationally
invariant, quadratic form in $m\left(r\right)$

$$\mathcal{H}_0 = \sum_{i=1}^{3} \frac{1}{V} \int \mathrm{d}^3r\,\mathrm{d}^3r'\,\Omega_i\left(r - r'\right) m_i\left(r\right) m_i\left(r'\right) \quad . \tag{20.4}$$

With this approach F_0 can be calculated and the a piori unknown function
$\Omega_i\left(r - r'\right)$ also serves as the variational parameter. Equation (20.4) has a
physically rather intuitive form since it describes the interaction of the ma-
gnetizations at points r and r' via a function $\Omega_i\left(r - r'\right)$ and thus strongly
reminds one on the Heisenberg Hamiltonian [see (7.1) in Chap. 7]. The following derivation is a basic example for the solution of the Landau–Ginzburg
Hamiltonian and is therefore given in detail. One starts by defining the Fourier transforms

$$m_i\left(r\right) = \sum_{k} m_{ki}\exp\left(\mathrm{i}kr\right) \quad , \quad \Omega_i\left(r\right) = \frac{1}{V}\sum_{k}\Omega_{ki}\exp\left(\mathrm{i}kr\right) \quad , (20.5)$$

which allow to write the trial Hamiltonian as

$$\mathcal{H}_0 = \sum_{k,i}\Omega_{ki} m_{ki} m_{-ki} = \sum_{k,i}\Omega_{ki}\left|m_{ki}\right|^2 \quad . \tag{20.6}$$

To calculate the respective sum of states Z_0 one has to integrate over the
phase space of the fluctuations which is given by the product over all independent variables

$$\mathrm{d}\Gamma = \prod_{k,i}\mathrm{d}m_{ki} \quad , \tag{20.7}$$

and hence

$$Z_0 = \prod_{\boldsymbol{k},i} \int\limits_{-\infty}^{+\infty} d\boldsymbol{m}_{\boldsymbol{k}i} \exp\left(-\beta \sum_{\boldsymbol{k},i} \Omega_{\boldsymbol{k}i} |\boldsymbol{m}_{\boldsymbol{k}i}|^2\right)$$

$$= \prod_{\boldsymbol{k},i} \left(\frac{\pi k_{\mathrm{B}} T}{\Omega_{\boldsymbol{k}i}}\right)^{\frac{1}{2}} , \tag{20.8}$$

so that the free energy F_0 becomes

$$F_0 = -k_{\mathrm{B}} T \ln Z$$

$$= -\frac{k_{\mathrm{B}} T}{2} \sum_{\boldsymbol{k},i} \ln\left(\frac{\pi k_{\mathrm{B}} T}{\Omega_{\boldsymbol{k}i}}\right) . \tag{20.9}$$

The Fourier transforms of the averages are

$$\left\langle m_i (r)^2\right\rangle = \sum_{\boldsymbol{k}_1, \boldsymbol{k}_2} \langle m_{\boldsymbol{k}_1 i} m_{\boldsymbol{k}_2 i}\rangle \exp\left[-i r \left(\boldsymbol{k}_1 + \boldsymbol{k}_2\right)\right]$$

$$= \sum_{\boldsymbol{k}} \left\langle |\boldsymbol{m}_{\boldsymbol{k}i}|^2\right\rangle . \tag{20.10}$$

The simplification of the summation to involve only one variable $\boldsymbol{k}$ is due to the translational symmetry. The thermodynamical average of the fluctuation with respect to $\mathcal{H}_0$ is given by

$$\left\langle m_i (r)^2\right\rangle = \sum_{\boldsymbol{k}} \frac{\int\limits_{-\infty}^{+\infty} |\boldsymbol{m}_{\boldsymbol{k}i}|^2 \exp\left(-\beta\mathcal{H}_0\right) d\Gamma}{\int\limits_{-\infty}^{+\infty} \exp\left(-\beta\mathcal{H}_0\right) d\Gamma}$$

$$= \sum_{\boldsymbol{k}} \frac{\int\limits_{-\infty}^{+\infty} d\boldsymbol{m}_{\boldsymbol{k}i} |\boldsymbol{m}_{\boldsymbol{k}i}|^2 \exp\left(-\beta\mathcal{H}_0\right) d\Gamma}{\int\limits_{-\infty}^{+\infty} d\boldsymbol{m}_{\boldsymbol{k}i} \exp\left(-\beta\mathcal{H}_0\right) d\Gamma}$$

$$= \frac{k_{\mathrm{B}} T}{2} \sum_{\boldsymbol{k}} \frac{1}{\Omega_{\boldsymbol{k}i}} . \tag{20.11}$$

The thermodynamical average of $\mathcal{H}_0$ is obtained along the same lines

$$\langle \mathcal{H}_0\rangle_0 = \sum_{\boldsymbol{k}i} \Omega_{\boldsymbol{k}i} \left\langle |\boldsymbol{m}_{\boldsymbol{k}i}|^2\right\rangle = \frac{k_{\mathrm{B}} T}{2} \sum_{\boldsymbol{k}i} 1 , \tag{20.12}$$

giving a contribution which is just proportional to the number of modes (= degrees of freedom). The final step is the calculation of the thermodynamical average of $\mathcal{H}$. One writes $\langle \mathcal{H}\rangle$ in the form

$$\langle \mathcal{H}\rangle = E(M) + \varphi + \frac{C}{2V} \int d^3 r \sum_{i,j} \langle \nabla_j m_i (r)\rangle^2 . \tag{20.13}$$

In (20.12) the function φ has been introduced which is the difference between the Landau expansion at $T = 0$ $E(M)$ and the respective expansion which contains the fluctuations for $T > 0$. It is given by

$$\varphi = \frac{1}{V} \int d^3r \left\langle E\left(M + \sum_{i=1}^{3} m_i(r)\right) - E(M)\right\rangle \quad . \tag{20.14}$$

Using Fourier transforms one can calculate the integral in (20.13) (see Sect. F.) and obtain

$$\langle \mathcal{H} \rangle_0 = E(M) + \varphi + \frac{C}{2}\frac{k_B T}{2} \sum_{ki} \frac{k^2}{\Omega_{ki}} \quad . \tag{20.15}$$

Now one can collect all terms and formulate the free energy from the Bogoliubov inequality

$$F \leq F_0 + \langle \mathcal{H} - \mathcal{H}_0 \rangle_0$$

$$= E(M) + \varphi + \frac{C}{2}\frac{k_B T}{2} \sum_{ki} \frac{k^2}{\Omega_{ki}}$$

$$- \frac{k_B T}{2} \sum_{ki} \left(1 + \ln\left(\frac{\pi k_B T}{\Omega_{ki}}\right)\right) \tag{20.16}$$

The function Ω_{ki} was introduced as a variational parameter so as to construct an optimal approximation for F. The value of Ω_{ki} is now determined from the condition $\frac{dF}{d\Omega_{ki}} = 0$ which yields

$$\Omega_{ki} = \frac{C}{2}k^2 + \frac{\partial \varphi}{\partial \langle m_i(r)^2 \rangle} \quad . \tag{20.17}$$

Up to 4th order in the magnetic moment the function φ is given

$$\varphi = \frac{A}{2}\left(\langle m_1^2 \rangle + \langle m_2^2 \rangle + \langle m_3^2 \rangle\right)$$

$$+ \frac{B}{4}\Big[M^2\left(2\langle m_1^2 \rangle + 2\langle m_2^2 \rangle + 6\langle m_3^2 \rangle\right) + 4\langle m_1^2 \rangle^2$$

$$+ 4\langle m_2^2 \rangle^2 + 3\langle m_3^2 \rangle^2 + 2\langle m_1^2 \rangle \langle m_3^2 \rangle + 2\langle m_2^2 \rangle \langle m_3^2 \rangle\Big], \tag{20.18}$$

where the obvious r dependence of $\langle m_i^2 \rangle$ was omitted. As in Sect. 18.2 an isotropic system is assumed so that $\langle m_1^2 \rangle = \langle m_2^2 \rangle = \langle m_\perp^2 \rangle$ and $\langle m_3^2 \rangle = \langle m_\parallel^2 \rangle$. Calculating the derivative of φ with respect to the fluctuations one obtains

$$\frac{\partial \varphi}{\partial \langle m_1^2 \rangle} = \frac{A}{2} + \frac{B}{4}\left(2M^2 + 8\langle m_1^2 \rangle + 2\langle m_3^2 \rangle\right) = \frac{1}{\chi_\perp} - \frac{C}{2}k^2 \quad , \tag{20.19}$$

for the perpendicular component, and

$$\frac{\partial \varphi}{\partial \langle m_3^2 \rangle} = \frac{A}{2} + \frac{B}{4} \left(6M^2 + 2 \langle m_1^2 \rangle + 2 \langle m_2^2 \rangle + 6 \langle m_3^2 \rangle \right) = \frac{1}{\chi_\parallel} - \frac{C}{2} k^2 \quad,$$

for the parallel component where one can interpret the result in terms of a perpendicular and parallel susceptibility $\chi_\perp$ and $\chi_\parallel$. It is easy to prove that both derivatives vanish at T_c for $k = 0$ which is the condition for a ferromagnetic Curie temperature. Of more interest is their value at $T = 0$ and $k = 0$. Here the perpendicular component $\frac{1}{\chi_\perp}$ is zero but the parallel component has the value of the inverse bulk susceptibility which is $(2\chi_0)^{-1}$. One can thus interpret these relations as susceptibilities for the excitation of parallel or perpendicular fluctuations. On closer inspection one finds that the perpendicular susceptibility is much larger (for finite k) than the parallel one, which means that it is much easier to excite a tilting of the moments than a fluctuation along the parallel component. This behavior again resembles the result for the Heisenberg model where also the tilting of the localized spins was found to be the important mechanism.

Combining (20.11) and (20.17) allows one to calculate the explicit form for the fluctuation amplitude

$$\langle m_i^2 \rangle = \frac{k_B T}{2} \frac{V}{8\pi^3} \int_0^{k_c} \frac{4\pi k^2}{\frac{C}{2} k^2 + \frac{\partial \varphi}{\partial \langle m_i(r)^2 \rangle}} dk \quad . \tag{20.20}$$

In (20.20) there appear two unknown parameters: the "spin wave stiffness" C and the cut-off wave vector k_c. The latter quantity must be introduced, since the Ornstein–Zernicke correlation function, which appears in the integrand, leads to a divergence if the integration is carried out to $|k| \to \infty$. A physical justification for the introduction of this cut-off may be found in the strong damping of the spin fluctuations by the single particle excitations as discussed in Sect. 19. From the zero temperature properties of the derivative of in the integrand of (20.20) one can replace the spin wave stiffness constant C by ξ^2/χ_0 where ξ is the correlation length and χ_0 is the susceptibility. Equation (20.20) yields only an implicit definition of $\langle m_i^2 \rangle$ because the fluctuation amplitude occurs also in the denominator. For practical application (20.20) must be solved numerically. Carrying out the integration yields

$$\langle m_i^2 \rangle = \frac{k_B T V k_c}{2\pi^2 C} \left[1 - \sqrt{\left(\frac{2 \frac{\partial \varphi}{\partial \langle m_i \rangle}}{C k_c^2} \right)} \arctan \sqrt{\left(\frac{C k_c^2}{2 \frac{\partial \varphi}{\partial \langle m_i \rangle}} \right)} \right] , \tag{20.21}$$

which together with the magnetic equation of state

$$H = \left(\frac{\partial E}{\partial M} \right) + \left(\frac{\partial \varphi}{\partial M} \right)_{T, m_i} \quad, \tag{20.22}$$

forms the set of equations to be solved self-consistently to determine the bulk moment M and the fluctuations $\langle m_\perp^2 \rangle$ and $\langle m_\parallel^2 \rangle$ at a given temperature T.

The indices to the derivatives on the rhs of (20.22) indicate that T and m_i have to be kept fixed; H is the (external) field.

From (20.16) it is straightforward to derive expressions for the entropy S and the specific heat c_H (at constant field H)

$$S = \frac{k_B}{2} \sum_{ki} \left(1 + \ln \frac{\pi k_B T}{2 \Omega_{ki}} \right) , \qquad (20.23)$$

and

$$c_H = \frac{k_B}{2} \sum_{ki} 1 - \sum_{ki} \langle m_i^2 \rangle \left(\frac{\partial^2 \varphi}{\partial T \partial \langle m_i^2 \rangle} \right)_H . \qquad (20.24)$$

The first term of the specific heat corresponds to the classical Dulong–Petit law, since it simply sums up the number of independent modes of the spin fluctuations. The second term gives the actual contribution due to the spin fluctuations. Its form is very similar to the one derived originally by Murata and Doniach [44]. For $T > T_c$ it becomes negative as can be seen from (18.27). It is easy to prove that for T= 0K the entropy diverges logarithmically. This violation of the 3rd law of thermodynamics is due to the neglect of quantum fluctuations. By introducing the effects of zero-point fluctuations an attempt was made to account for this shortcoming [197].

The properties of the derivatives of the function φ allow one to derive an expression for the Curie temperature of a spin fluctuation system. Since $\frac{\partial \varphi}{\partial \langle m_i(r)^2 \rangle}$ must vanish at $T = T_c$ one can carry out the integration at T_c and obtains a result similar to the Mohn–Wohlfarth model

$$T_c = \frac{M_0^2}{10 \chi_0 k_B} \frac{4 \pi^2 \xi^2}{V k_c} . \qquad (20.25)$$

In contrast to the long wavelength expression (18.24), (20.25) also contains the missing information about the finite wavelength dependence of the fluctuations which should lead to a wider applicability.

It should be noted that the integrand of (20.20) can be interpreted as the wavevector dependent susceptibility. The integrand is essentially given by (20.17) and reads

$$\chi_0(k) = \frac{1}{\Omega_{k\parallel}} = \frac{1}{\frac{C}{2} k^2 + \frac{1}{2\chi_0}}$$

$$= \frac{2\chi_0}{1 + \xi^2 k^2} , \qquad (20.26)$$

where the $T = 0$ value for $\frac{\partial \varphi}{\partial \langle m_\parallel(r)^2 \rangle}$ was used. ξ is again the correlation length which is also sometimes called "the real space parameter". The role of ξ becomes clear from Fig. 20.1, which shows that $\frac{1}{\xi}$ determines the halfwidth of the k-distribution.

The function $\chi(k)$ is of Lorentz shape, where the correlation length ξ determines the width of the distribution. The solution given in (20.26) above

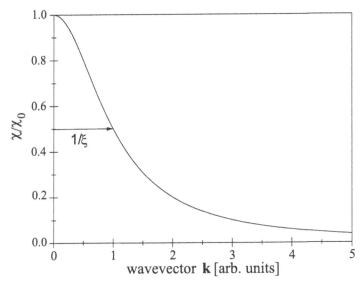

Fig. 20.1. Wavevector dependent susceptibility of a ferromagnetic fluctuation system

is written for low temperatures because ξ also shows a temperature dependence which causes a broadening of the distribution at higher temperature (at $T = T_c$ the correlation length actually diverges). The simple Ornstein–Zernicke form of $\chi(\boldsymbol{k})$ arises from the assumption of a quadratic form for $\mathcal{H}_0$, however, it is found that most ferromagnetic systems are quite well described.

How can one interpret this form for the susceptibility? The susceptibility in its most general form describes the answer of a system to a perturbation. This perturbation can be almost anything (e.g. an electric field or as here our case a magnetic one). When the susceptibility was introduced from the beginning a static homogenous field H was assumed. The response of the system was a static, homogenous magnetization M. If one now applies a field which varies in space with a wavevector $\boldsymbol{k}$ (this field drives the local fluctuations) the answer is $\chi(\boldsymbol{k})$. A mechanical example is the forced oscillation of a pendulum. The maximum of $\chi(\boldsymbol{k})$ for $k = 0$ is nothing more but the resonance frequency (ferromagnetic resonance) and for any deviation from the resonance the amplitude becomes smaller and follows exactly the Lorentz curve. But one can take this analogue even further considering two coupled pendulums. There are two frequencies one for the uniform mode (the pendulums are in phase: ferromagnetism) and one for the antiphase mode (antiferromagnetism). This also explains why there is no Stoner criterion for antiferromagnetism, because the instability towards antiferromagnetic order can only be derived from the dynamical properties of the system and not from its non-magnetic ground state.

In general Landau–Ginzburg theory of spin fluctuations yields fairly good results. The deviation of the Curie temperatures and the susceptibilities are within 15 % as compared to experiment. This means that spin fluctuations cover the essential physics of magnetic coupling. The order of the magnetic phase transitions is less well described. The Landau–Ginzburg theory shows a tendency to produce first order phase transitions. In his review Shimizu [207] blames this tendency on the use of of the Gaussian fluctuations. But a detailed understanding of these effects is not yet at hand.

21. Conclusion and Lookout

This book has attempted to review the most common models of solid state magnetism. At the present stage of development the spin fluctuation theory is the most promising phenomenological model. In the form used nowadays it combines the quantum mechanical ground state properties which are well described by the Stoner model with a classical treatment of collective excitations of the spins of the band electrons. These excitations can already be excited at low temperature and also persist above T_c. The paramagnetic state for $T > T_c$ is no longer the non-magnetic state as in the Stoner model, but it is a state where the correlation between the spins has broken down (no long range order) but the individual moments still exist. With spin fluctuation theory at hand, one can complete the analogy between the localized and itinerant moment systems:

For localized moments and sufficiently high temperatures one can apply the Weiss model, which considers only the z-component of the angular momentum (spin) neglecting all quantum fluctuations. The Curie temperature is given by the temperature where all allowed states with quantum number m_j are equally occupied so that the net magnetic moment is zero. The Heisenberg model corrects for the omitted of quantum fluctuations by describing also a tilting of the localized spins produced by collective excitations. Since collective excitations involve more than one electron (spin) the Heisenberg model goes beyond the single particle picture. At the Curie temperature there is no complete disorder, but only the long range correlation between the spins of the crystal breaks down. Collective excitations (magnons) also exist above T_c.

For itinerant moments one starts from the Stoner model, which again considers only the z-component of the spin of the band-electrons. The Curie temperature is given by the temperature where single particle excitations have reduced the band splitting to zero so that the subbands for spin-up and spin-down are equally occupied. The paramagnetic state above T_c is described as the non-magnetic state and the resulting Curie temperature is unrealistically high. An improvement, which is beyond the scope of this book, can be formulated within the "random phase approximation" (RPA) or the "dynamical Hartree–Fock" theory which also allows one to formulate spinwave excitations. Within this theory the low temperature behavior of itinerant systems can be described rather well. At high temperatures, however, the RPA fails, since the fluctuations are calculated only with respect to the $(T = 0K)$ ground state. An improvement is provided by the "self-consistent renormalization theory" for the spin fluctuations (SCR), where the renormalization accounts for corrections in the free energy which a due to the thermal excitations of the fluctuations [43]. The Landau–Ginzburg theory of spin fluctuations presented in this book can be seen as a phenomenological formulation of Moriya's SCR theory. Spin fluctuations correct for this shortcoming by allowing for fluctuations parallel and perpendicular to the bulk magnetic moment. These fluctuations are again collective modes which consist of local randomly fluctuating moments which are induced by temperature. At the Curie temperature the bulk magnetic moment vanishes since the fluctuations destroy the long range correlations between the band electrons. However the band splitting and the fluctuations persist above T_c so that the paramagnetic state is no longer the non-magnetic state. A review of itinerant electron magnetism which also includes spin fluctuations was given by Shimizu [207].

Going beyond the aim of this course, namely to introduce solid state magnetism on the basis of the extreme cases of localization and itineracy, there exist a number of promising efforts to combine these concepts in order to formulate a unified model. The most advanced concept for the description of spin fluctuations is the "functional integral method" [43, 183, 184] which allows one to describe both localized and itinerant systems. Calculations performed on the basis of this theory yield quite a good description of the thermodynamical properties of the magnetic quantities [185]–[188]. The disadvantage of the functional integral method is that it can only solved analytically for highly idealized densities of states so that for real systems analytical results cannot be obtained.

A different track that has been followed started with the first self-consistent spin-polarized electronic band structure calculations which emerged in the mid-1970s. These calculations yielded satisfactory results for the magnetic moment, the cohesive energy and the stability of a magnetic ground state within the Stoner model which can be mapped to density functional theory. It was shown that the results of such band structure calculations can be mapped to a thermodynamical description of spin fluctuations as given in Chap. 18 [189]–[191]. A different method was devised by Nolting and coworkers who plugged the single particle band structure into a Hubbard type Hamiltonian [192]. With the parameters of the Hubbard model fitted to experiment, excellent results for the Curie temperature were obtained.

In the "disordered local moment" (DLM) approach, the magnetic moments on individual atoms are allowed to have random orientations in the sense described by Herring [193]. In a series of papers it was shown how the DLM picture can be implemented into a Korringa–Kohn–Rostoker (KKR) coherent potential approximation (CPA) band structure method [194, 196].

Within this DLM approach it was possible to calculate the wave vector dependent susceptibility [195] and *ab initio* Curie temperatures of 1260 K and 225 K for iron and nickel, respectively. Although the comparison with experiment is reasonably good (in particular for Fe), the remaining discrepancy is assumed to be caused by the fact that above T_c spin disorder is complete so that effects of short range order are neglected as pointed out already earlier [198]–[200]. The existence of local moments with in the DLM state above the Curie temperature, which is important for a quantitative understanding of the Curie constant, has been addressed by Heine et al. [201]. For bcc Fe they calculate a local moment above T_c of about $1.8\mu_B$ which is in agreement with the experimental placement of Fe close to the localized limit in the Rhodes–Wohlfarth plot (Fig. 6.5).

Another implementation of spin non-collinearity was introduced by Sandratskii [202]–[205] who formulated the so called spin-spiral approach. On the basis of these ab-initio band structure calculations of non-collinear spin-spirals, is was possible to derive the phenomenological parameters which appear in the Landau–Ginzburg theory of spin fluctuations from first principles. This approach formulated within the "exchange coupled" spin fluctuation model also yielded fair results for the ordering temperatures of Fe, Co, and Ni [206] and for the finite temperature properties of Invar alloys [208]. Both Rosengaard et al. [209] and Halilov et al. [210] follow similar lines by mapping the results of spin-spiral calculations to Heisenberg-type Hamiltonians.

In a slightly different approach, Antropov and coworkers [211, 212] formulated spin-dynamics within the density-functional theory in a way which allows one to use Monte-Carlo type simulations.

It is obvious that the possibility of employing massive computer power has also changed the approach to solid state magnetism. The present strategies are less devoted to achieving analytical solutions of microscopic models, but rather to large scale numerical simulations. However, the interpretation of the numerical results is still based on the understanding of the elementary models which have been presented in this book.

A. Appendices

A. Convexity Property of the Free Energy

A function $g(x)$ is a convex function of its argument x if the inequality holds

$$g\left(\frac{x_1 + x_2}{2}\right) \leq \frac{g(x_1) + g(x_2)}{2} \quad , \tag{A.1}$$

for all x_1 and x_2. This property implies that the second derivative of such a function must be ≥ 0. Applied to the free energy this condition means that

$$\left.\frac{\partial^2 F}{\partial T^2}\right|_H = -\frac{c_H}{T} \quad , \qquad \left.\frac{\partial^2 F}{\partial H^2}\right|_T = -\chi \quad . \tag{A.2}$$

where c_H is the specific heat and χ is the susceptibility. The 3rd law of thermodynamics demands that the specific heat must be positive. Also the susceptibility usually has a non-negative value (exceptions are diamagnetic susceptibilities). Because the second derivatives of of the free energy with respect to T and H are negative it is a concave function of both his variables. It should be noted that the convexity property is not a condition which must be fulfilled over the whole range of the free energy. Usually the free energy shows minima, maxima or even discontinuities which violate the convexity property locally. However, globally the free energy has to be a convex function for $x \to \pm\infty$ so that $F(+\infty) \to +\infty$ and $F(-\infty) \to +\infty$.

The condition formulated in (A.1) can also be written in a form which is more directly related to the statistical averaging process of thermodynamics. A function is convex if the following relation holds

$$g(\langle x \rangle) \leq \langle g(x) \rangle \quad , \tag{A.3}$$

where $\langle\rangle$ denotes the average in the same way as in (A.1).

B. Derivation of the Coefficient a in (3.17)

The coefficient a describes the temperature dependence of the paramagnetic susceptibility as given by (3.17). A similar derivation would have been necessary to obtain the coefficient b for the specific heat (2.49). At this point it will be shown how the "magnetic" entity a can be derived, a procedure which can also be applied to derive the specific heat coefficient b.

With the obvious property of the magnetic moment

$$M = \mu_B \left(n^+ - n^-\right) \quad ,$$

one introduces an enthalpy type of function $G(x)$ which allows one to write the magnetic moment as

$$M = \mu_B \left(G\left(\eta + \beta\right) - G\left(\eta - \beta\right)\right) \quad .$$

From the analogy between G and the number of particles $n^\pm$ one derives an explicit form for G

$$G(\eta) = \int_0^\infty \mathcal{N}(\varepsilon) \left(\exp\left(\frac{\varepsilon}{k_B T} - \eta\right) + 1\right)^{-1} d\varepsilon$$

$$= \int_0^\infty \mathcal{N}(\varepsilon) f(\varepsilon) \, d\varepsilon \quad .$$

For $\beta << 1$, $G(\eta + \beta)$ is expanded in a Taylor series:

$$G(\eta + \beta) = G(\eta) + \beta \frac{dG(\eta)}{d\eta} + \dots$$

$$\Rightarrow M = \mu_B \left(G(\eta) + \beta \frac{dG(\eta)}{d\eta} - G(\eta) + \beta \frac{dG(\eta)}{d\eta}\right)$$

$$= 2\mu_B^2 \beta \frac{dG(\eta)}{d\eta} = \frac{2\mu_B^2 H}{k_B T} \frac{dG(\eta)}{d\eta} \quad .$$

How does $\frac{dG(\eta)}{d\eta}$ look like ?

$$\frac{dG(\eta)}{d\eta} = \int_0^\infty \frac{d}{d\eta} \left(\mathcal{N}(\varepsilon) f(\varepsilon)\right) d\varepsilon$$

$$= \int_0^\infty \mathcal{N}(\varepsilon) \frac{\exp\left(\frac{\varepsilon}{k_B T} - \eta\right)}{\left(\exp\left(\frac{\varepsilon}{k_B T} - \eta\right) + 1\right)^2} d\varepsilon$$

$$= -k_B T \int_0^\infty \mathcal{N}(\varepsilon) \frac{df(\varepsilon)}{d\varepsilon} d\varepsilon$$

$$= k_B T \int_0^\infty \frac{d\mathcal{N}(\varepsilon)}{d\varepsilon} f(\varepsilon) \, d\varepsilon \quad \text{(by part. integr.)}$$

(during the last step the identity $\frac{dG}{d\varepsilon} = 0$ was used).

This allows one to write the susceptibility as

$$\chi = \frac{2\mu_B^2}{k_B T} k_B T \int_0^\infty \frac{\mathrm{d}\mathcal{N}(\varepsilon)}{\mathrm{d}\varepsilon} f(\varepsilon) \, \mathrm{d}\varepsilon$$

Then it follows from (2.38) and (2.39) that

$$\int_0^\infty \mathcal{N}(\varepsilon) f(\varepsilon) \, \mathrm{d}\varepsilon = \int_0^{\varepsilon_T} \mathrm{d}\mathcal{N}(\varepsilon) \, \mathrm{d}\varepsilon \quad .$$

By means of the Taylor series given above one is able to calculate the number of particles

$$\frac{n}{2} = \frac{1}{2}(G(\eta + \beta) + G(\eta - \beta)) = G(\eta) \quad ,$$

$$\frac{n}{2} = \int_0^{\varepsilon_T} \mathcal{N}(\varepsilon) \, \mathrm{d}\varepsilon + \frac{\pi^2}{6}(k_B T)^2 \mathcal{N}(\varepsilon_T)'$$

$$= \int_0^{\varepsilon_T} \mathcal{N}(\varepsilon) \, \mathrm{d}\varepsilon + (\varepsilon_T - \varepsilon_F) \mathcal{N}(\varepsilon_F) + \frac{\pi^2}{6}(k_B T)^2 \mathcal{N}(\varepsilon_F)' \quad ,$$

$$\Rightarrow \varepsilon_T = \varepsilon_F - \frac{\pi^2}{6}(k_B T)^2 \frac{\mathcal{N}(\varepsilon_F)'}{\mathcal{N}(\varepsilon_F)} \quad .$$

Using this expression for the "temperature dependence" of the Fermi energy one writes for the susceptibility

$$\frac{\chi}{2\mu_B^2} = \int_0^{\varepsilon_T} \frac{\mathrm{d}\mathcal{N}(\varepsilon)}{\mathrm{d}\varepsilon} \mathrm{d}\varepsilon + \frac{\pi^2}{6}(k_B T)^2 \mathcal{N}(\varepsilon_F)''$$

$$= \mathcal{N}(\varepsilon_F) - (\varepsilon_T - \varepsilon_F)\mathcal{N}(\varepsilon_F)' + \frac{\pi^2}{6}(k_B T)^2 \mathcal{N}(\varepsilon_F)''$$

$$= \mathcal{N}(\varepsilon_F) - \frac{\pi^2}{6}(k_B T)^2 \frac{(\mathcal{N}(\varepsilon_F)')^2}{\mathcal{N}(\varepsilon_F)} + \frac{\pi^2}{6}(k_B T)^2 \mathcal{N}(\varepsilon_F)'' \quad ,$$

$$\Rightarrow \chi = \chi_0 \left(1 + \frac{\pi^2}{6}(k_B T)^2 \left[\frac{\mathcal{N}(\varepsilon_F)''}{\mathcal{N}(\varepsilon_F)} - \left(\frac{\mathcal{N}(\varepsilon_F)'}{\mathcal{N}(\varepsilon_F)}\right)^2\right]\right) \quad \text{q. e. d.}$$

which is identical with the result given by (3.17).

C. Quenching of the Orbital Momentum

For a free atom the second Hund's rule requires that the orbital momentum always has to be a maximum. This means e.g. that for the $3d$ atom Ti the two d-electrons occupy the angular momentum states $m_l = +2$ and $m_l = +1$, and have parallel spin (Hund's first rule). This result is based on the spin-orbit interaction which, however small for the transition metals, lifts the degeneracy of the five d-levels; the relevant quantum number for the classification of these states is the total momentum J.

In a solid the major interaction stems from the neighboring atoms or ions an effect which gives rise to the so called crystal field splitting (in chemistry one rather uses the term ligand field splitting). For the $3d$-elements this splitting is much larger than the spin-orbit splitting, so that J is no longer a good quantum number and the latter effect can usually be treated as a small perturbation. The crystal field splitting, which is a "geometrical" effect is due to the electrostatic interaction among the ions or due to covalent interactions and thus causes a breaking of the full spherical symmetry of the free atom resembling the symmetry of the crystal. Since the cubic crystal structure is very common for the transition metals it will now be shown how the cubic crystal field quenches the orbital momentum.

If one considers the angular dependencies of the five d-wave functions only one can write them

$$
\begin{aligned}
\Psi_{\pm 2} &\sim \sin^2 \theta \exp\left(\pm i2\varphi\right) \quad, \\
\Psi_{\pm 1} &\sim \sin \theta \cos \theta \exp\left(\pm i\varphi\right) \quad, \\
\Psi_0 &\sim \left(3\cos \theta - 1\right) \quad.
\end{aligned} \tag{C.1}
$$

These functions can be rewritten as linear combinations to obtain a different basis which fulfills the Hamiltonian but which are now no longer imaginary functions. One obtains the so called real spherical harmonics

$$
\begin{aligned}
K_0 &\equiv \Psi_0 \sim \frac{\left[2z^2 - x^2 - y^2\right]}{r^2} \quad, \\
K_{2+} &\equiv \frac{1}{\sqrt{2}}\left[\Psi_{+2} + \Psi_{-2}\right] \sim \frac{\left[x^2 - y^2\right]}{r^2} \quad, \\
K_{2-} &\equiv \frac{1}{\sqrt{2}}\left[\Psi_{+2} - \Psi_{-2}\right] \sim \frac{\left[xy\right]}{r^2} \quad, \\
K_{1+} &\equiv \frac{1}{\sqrt{2}}\left[\Psi_{+1} + \Psi_{-1}\right] \sim \frac{\left[zx\right]}{r^2} \quad, \\
K_{1-} &\equiv \frac{1}{\sqrt{2}}\left[\Psi_{+1} - \Psi_{-1}\right] \sim \frac{\left[yz\right]}{r^2} \quad.
\end{aligned} \tag{C.2}
$$

It can be shown that these real spherical harmonics are the eigenfunctions of the Hamiltonian of an atom in a cubic crystal field. One finds that these originally degenerate five states split into the 3-fold degenerate Γ_5 state (t_{2g}) which contains K_{2-}, K_{1+}, and K_{1-} and into the 2-fold degenerate Γ_3 state (e_g) which contains K_0 and K_{2+}. Since the eigenfunctions of the atom in a cubic crystal field are real functions the expectation value of the angular momentum operator $\mathbf{L}$ (which is complex) with regard to these eigenfunctions is zero. Or less formal, since any of these states combines a wavefunction with $+m$ and $-m$ the expectation value for $\mathbf{L}$ is $+m$ and $-m$ at the same time, which can only be fulfilled if m is zero. This means that the orbital momentum is quenched or at least very small, since also for these states one observes spin-orbit interaction which recreates a finite orbital momentum.

Going beyond this "hand waving" explanation of the quenching of the orbital momentum this effect can be derived completely general from the properties of the angular momentum operator concerning time reversal symmetry [178]. Since the operator $\mathbf{L}$ contains a generalized velocity, time reversal transforms $\mathbf{L}$ into $-\mathbf{L}$. Let $\mathbf{T}$ be the time reversal operator with the property that $\mathbf{T}^{-1}\mathbf{T} = 1$ (the unity operator) it is easy to see that since it acts only on the angular momentum part of the wave function (C.1) one can write

$$\mathbf{T}\Psi = \Psi^* \quad . \tag{C.3}$$

Since the Hamiltonian is hermitian Ψ and Ψ^* have the same eigenvalue. If now Ψ is the wave function of a non-degenerate state Ψ and Ψ^* must be linearly dependent, $\Psi^* = c\Psi$ and by applying $\mathbf{T}$ to this relation one finds that in particular $\Psi = |c|^2 \Psi$. This means that c can be written as $\exp(i\varphi)$ where φ is real. One can now calculate the expectation value of $\mathbf{L}$ between two states $\langle n|$ and $|m\rangle$

$$\langle n| \mathbf{L} |m\rangle = \langle n| \mathbf{T}^{-1}\mathbf{T}\mathbf{L}\mathbf{T}^{-1}\mathbf{T} |m\rangle$$
$$= -\langle n| \mathbf{T}^{-1}\mathbf{L}\mathbf{T} |m\rangle = -\exp i(\varphi_n - \varphi_m) \langle n| \mathbf{L} |m\rangle^* \quad . \tag{C.4}$$

The expectation value of $\mathbf{L}$ for a non-degenerate eigenstate $|n\rangle$ is given by

$$\langle n| \mathbf{L} |n\rangle = -1 \langle n| \mathbf{L} |n\rangle^* \quad . \tag{C.5}$$

Relation (C.5) can only be fulfilled if the expectation value calculated is pure imaginary. Since $\mathbf{L}$ measures an observable it must be real and thus zero.

One can formulate the general theorem: i) The matrix element of the orbital angular momentum between non-degenerate states has an arbitrary phase, which in particular may be pure real or pure imaginary. ii) The expectation value of $\mathbf{L}$ for any non-degenerate state is zero.

If the crystal field now lifts the degeneracy of the atomic states it quenches the orbital momentum.

D. Properties of "Classical" Spins

A less intuitive solution of our model of classical spins shall be given in the following paragraph. Consider a system of N magnetic moments "classical spins" which have two states $\mu = \pm Sg\mu_B$ ($S = \frac{1}{2}$, $g = -2$). These two states are given by magnetization parallel or antiparallel to an applied field H so that the respective change in energy becomes $\pm\mu H$. The sum of states for such a system is given by

$$Z = \left[\exp\left(+\frac{\mu H}{k_B T}\right) + \exp\left(-\frac{\mu H}{k_B T}\right)\right]^N = \left[2\cosh\left(\frac{\mu H}{k_B T}\right)\right]^N . \quad \text{(D.1)}$$

From the sum of states one can calculate all properties of this system. The relevant relations are

$$
\begin{aligned}
\text{Free energy} \quad & F = -k_B T \ln Z \quad, \\
\text{Internal energy} \quad & U = -\frac{\partial \ln Z}{\partial \beta} \quad, \\
\text{Entropy} \quad & S = -\left(\frac{\partial F}{\partial T}\right)_H = (U - F)/T \quad, \quad \text{(D.2)} \\
\text{Magnetization} \quad & M = -\left(\frac{\partial F}{\partial H}\right)_T \quad, \\
\text{Specific heat} \quad & c_x = T\left(\frac{\partial S}{\partial T}\right)_{x=H,M} \quad, \\
\text{Isothermal susceptibility} \quad & \chi_T = \left(\frac{\partial M}{\partial H}\right)_T \quad.
\end{aligned}
$$

In our case this yields for the free energy

$$F = -k_B T \ln Z = -k_B T N \left(\ln 2 + \ln\cosh\left(\frac{\mu H}{k_B T}\right)\right) \quad, \quad \text{(D.3)}$$

and the magnetization

$$
\frac{M}{N} = m = -\left(\frac{\partial F}{\partial H}\right) = \frac{\partial}{\partial H}(k_B T \ln Z) = k_B T \frac{1}{Z}\frac{\partial Z}{\partial H}
$$

$$
= \mu\frac{\exp\left(+\frac{\mu H}{k_B T}\right) - \exp\left(-\frac{\mu H}{k_B T}\right)}{\exp\left(+\frac{\mu H}{k_B T}\right) + \exp\left(-\frac{\mu H}{k_B T}\right)} = \mu\tanh\left(\frac{\mu H}{k_B T}\right). \quad \text{(D.4)}
$$

The second line of (D.4) can easily be interpreted insofar as the expression measures the probability of states parallel (with $+\mu$) minus the probability of states antiparallel (with $-\mu$) to the applied field. In terms of fractional occupation numbers this would read: $m = \mu(n^+ - n^-)$. For $T \to 0$ the tanh $\to 1$, so that $m \to \mu$. For zero temperature an infinitesimal magnetic field aligns all spins of the system.

The entropy is given as

$$\frac{S}{N} = -\left(\frac{\partial F}{\partial T}\right)_H$$

$$= k_B \left(\ln 2 + \ln\left[\cosh\left(\frac{\mu H}{k_B T}\right) - \frac{\mu H}{k_B T} \tanh\left(\frac{\mu H}{k_B T}\right)\right]\right) . \qquad (\text{D.5})$$

For large temperature the entropy takes the value $k_B \ln 2$ which is the classical result for a system with 2 degrees of freedom. However, for $T \to 0$ the entropy diverges logarithmically to $+\infty$. This divergence and subsequent violation of the 3rd law of thermodynamics is caused by our classical model which neglects quantum effects. Defining our model it was assumed that these two states have an exact energy and thus, at $T = 0$, an infinite lifetime, which is a violation of the uncertainty relation. To account for this problem one must replace our classical spin S by its quantum mechanical operator equivalent. In this case one could of course measure the z-component of the spin, but its x- and y-components remain uncertain. Due to the fact that the entropy diverges only logarithmically, the term TS in the internal energy always remains finite even at $T = 0$ where it becomes 0 as well.

The specific heat becomes

$$\frac{c}{N} = T\frac{\partial S}{\partial T} = \frac{\mu^2 H^2}{k_B T^2} \frac{1}{\cosh^2\left(\frac{\mu H}{k_B T}\right)} . \qquad (\text{D.6})$$

This function is the classical result for the specific heat of a two-level system. For low temperature it rises exponentially, goes through a maximum and for high temperature it approaches zero like $\frac{1}{T^2}$. Such a behavior is often called a Schottky-anomaly.

For the susceptibility one finally gets

$$\frac{\chi}{N} = \frac{\partial M}{\partial H} = \frac{\mu^2}{k_B T} \frac{1}{\cosh^2\left(\frac{\mu H}{k_B T}\right)} . \qquad (\text{D.7})$$

One immediately notices the close relation between the specific heat and the susceptibility. This is not surprising since both functions are response functions which describe the answer of the system to a perturbation either the temperature or the magnetic field.

For large temperature the susceptibility becomes proportional to $\frac{1}{T}$ and thus shows the Curie–Weiss behavior. For $T \to 0$ one has to distinguish between the case of a finite field H where the susceptibility is always 0 for $T = 0$. For $H = 0$ the susceptibility diverges to $+\infty$ which means that for $T = 0$ an infinitesimally small H field causes a full polarization.

Again one observes that a classical system of non-interacting particles shows a strange low-temperature behavior. Only a quantum-mechanical description can account for the true behavior.

E. Derivation of the Constant c in (8.24)

The constant c is very similar to the constant b involved in the temperature dependence of the paramagnetic susceptibility. For its derivation one again applies the formalism of the Sommerfeld expansion. To derive c one writes the spin up and spin down Fermi energies as

$$\varepsilon^+ = \varepsilon_F + \delta^+ \quad \text{and} \quad \varepsilon^- = \varepsilon_F - \delta^- \ ,$$

and expands the density of states in powers of $\delta^\pm$

$$\mathcal{N}(\varepsilon^+) = \mathcal{N}(\varepsilon_F) + \delta^+ \mathcal{N}(\varepsilon_F)' + \frac{1}{2}(\delta^+)^2 \mathcal{N}(\varepsilon_F)'' + \dots$$

$$= \mathcal{N}(\varepsilon_F)\left(1 + \delta^+ \frac{\mathcal{N}(\varepsilon_F)'}{\mathcal{N}(\varepsilon_F)} + \frac{1}{2}(\delta^+)^2 \frac{\mathcal{N}(\varepsilon_F)''}{\mathcal{N}(\varepsilon_F)} + \dots\right), \quad (\text{E.1})$$

and analogous for $\mathcal{N}(\varepsilon^-)$. This in turn gives for the reciprocal density of states

$$\frac{1}{\mathcal{N}(\varepsilon^+)} = \frac{1}{\mathcal{N}(\varepsilon_F)}$$

$$\times \left[1 - \delta^+ \frac{\mathcal{N}(\varepsilon_F)'}{\mathcal{N}(\varepsilon_F)} - \frac{1}{2}(\delta^+)^2 \frac{\mathcal{N}(\varepsilon_F)''}{\mathcal{N}(\varepsilon_F)} + (\delta^+)^2 \left(\frac{\mathcal{N}(\varepsilon_F)'}{\mathcal{N}(\varepsilon_F)}\right)^2 + \dots\right],$$

$$(\text{E.2})$$

$$\frac{1}{\mathcal{N}(\varepsilon^-)} = \frac{1}{\mathcal{N}(\varepsilon_F)}$$

$$\times \left[1 + \delta^- \frac{\mathcal{N}(\varepsilon_F)'}{\mathcal{N}(\varepsilon_F)} - \frac{1}{2}(\delta^-)^2 \frac{\mathcal{N}(\varepsilon_F)''}{\mathcal{N}(\varepsilon_F)} + (\delta^-)^2 \left(\frac{\mathcal{N}(\varepsilon_F)'}{\mathcal{N}(\varepsilon_F)}\right)^2 + \dots\right].$$

$$(\text{E.3})$$

Adding up these two terms yields

$$\frac{1}{\mathcal{N}(\varepsilon_F)}\left[2 - (\delta^+ - \delta^-)\frac{\mathcal{N}(\varepsilon_F)'}{\mathcal{N}(\varepsilon_F)}\right.$$

$$\left. -\frac{1}{2}\left((\delta^+)^2 + (\delta^-)^2\right)\left(\frac{\mathcal{N}(\varepsilon_F)''}{\mathcal{N}(\varepsilon_F)} - 2\left(\frac{\mathcal{N}(\varepsilon_F)'}{\mathcal{N}(\varepsilon_F)}\right)^2\right)\right]. \quad (\text{E.4})$$

The respective magnetic moment for one spin direction is given by

$$\frac{1}{2}n\zeta = \int_{\varepsilon_F}^{\varepsilon_F + \delta^+} \mathcal{N}(\varepsilon)\, d\varepsilon$$

$$= \int_{\varepsilon_F}^{\varepsilon_F + \delta^+} \left[\mathcal{N}(\varepsilon_F) + (\varepsilon - \varepsilon_F)\mathcal{N}(\varepsilon_F)' + \frac{1}{2}(\varepsilon - \varepsilon_F)^2 \mathcal{N}(\varepsilon_F)''\right] d\varepsilon$$

$$= \delta^{+} \mathcal{N}\left(\varepsilon_{\mathrm{F}}\right) + \frac{1}{2}\left(\delta^{+}\right)^{2} \mathcal{N}\left(\varepsilon_{\mathrm{F}}\right)' \quad , \tag{E.5}$$

and analogously for ε^{-} and δ^{-}

$$\frac{1}{2} n\zeta = \delta^{-} \mathcal{N}\left(\varepsilon_{\mathrm{F}}\right) + \frac{1}{2}\left(\delta^{-}\right)^{2} \mathcal{N}\left(\varepsilon_{\mathrm{F}}\right)' \quad . \tag{E.6}$$

Since $\left(\delta^{-}\right)^{2}$ can be expressed as

$$\left(\delta^{-}\right)^{2} = \left(\varepsilon - \varepsilon_{\mathrm{F}}\right)^{2} = \frac{n^{2}\zeta^{2}}{4\mathcal{N}\left(\varepsilon_{\mathrm{F}}\right)^{2}} \quad ,$$

one obtains for δ^{+} and δ^{-}

$$\delta^{+} = \frac{1}{2} \frac{n\zeta}{\mathcal{N}\left(\varepsilon_{\mathrm{F}}\right)} - \frac{1}{8} \frac{n^{2}\zeta^{2}}{\mathcal{N}\left(\varepsilon_{\mathrm{F}}\right)^{2}} \frac{\mathcal{N}\left(\varepsilon_{\mathrm{F}}\right)'}{\mathcal{N}\left(\varepsilon_{\mathrm{F}}\right)} \quad , \tag{E.7}$$

$$\delta^{-} = \frac{1}{2} \frac{n\zeta}{\mathcal{N}\left(\varepsilon_{\mathrm{F}}\right)} + \frac{1}{8} \frac{n^{2}\zeta^{2}}{\mathcal{N}\left(\varepsilon_{\mathrm{F}}\right)^{2}} \frac{\mathcal{N}\left(\varepsilon_{\mathrm{F}}\right)'}{\mathcal{N}\left(\varepsilon_{\mathrm{F}}\right)} \quad , \tag{E.8}$$

and finally

$$\frac{1}{\mathcal{N}\left(\varepsilon^{+}\right)} + \frac{1}{\mathcal{N}\left(\varepsilon^{-}\right)} \tag{E.9}$$

$$= \frac{2}{\mathcal{N}\left(\varepsilon_{\mathrm{F}}\right)} \left[1 - \frac{1}{8} \frac{n^{2}\zeta^{2}}{\mathcal{N}\left(\varepsilon_{\mathrm{F}}\right)^{2}} \left(\frac{\mathcal{N}\left(\varepsilon_{\mathrm{F}}\right)''}{\mathcal{N}\left(\varepsilon_{\mathrm{F}}\right)} - 3 \left(\frac{\mathcal{N}\left(\varepsilon_{\mathrm{F}}\right)'}{\mathcal{N}\left(\varepsilon_{\mathrm{F}}\right)} \right)^{2} \right) + \dots \right]$$

$$= \frac{2}{\mathcal{N}\left(\varepsilon_{\mathrm{F}}\right)} \left(1 - c\zeta^{2} \right) \quad . \tag{E.10}$$

F. Ornstein–Zernicke Extension

A useful example is provided by the Ornstein–Zernicke extension to Landau theory (The results of this section will be used for the derivation of spin-fluctuation theory). It should be noted that this model is also often called "Landau–Ginzburg" model or "continuum Gaussian" model. The free energy reads

$$F - F_{0} = a_{2}' t \int \left[m\left(\boldsymbol{r}\right) \right]^{2} \mathrm{d}^{3}r + g \int \left[\nabla m\left(\boldsymbol{r}\right) \right]^{2} \mathrm{d}^{3}r \quad . \tag{F.1}$$

The first term on the rhs is just the quadratic term in (14.1) written as an integral over 3-dimensional space to allow for the $\boldsymbol{r}$ dependence of the magnetization. The second term can be interpreted as the lowest order term in the expansion of the spin-spin interaction and takes into account the extra free energy that results if the spins are not parallel. This latter interpretation will become obvious when one rewrites it in its Fourier transformed form. The Fourier transform of the magnetization is

$$m\left(\boldsymbol{r}\right) = \frac{1}{\left(2\pi\right)^3} \int m\left(\boldsymbol{q}\right) \exp\left(\mathrm{i}\boldsymbol{q}\boldsymbol{r}\right) \mathrm{d}^3 q \quad , \tag{F.2}$$

$$m\left(\boldsymbol{q}\right) = \int m\left(\boldsymbol{r}\right) \exp\left(-\mathrm{i}\boldsymbol{q}\boldsymbol{r}\right) \mathrm{d}^3 r \quad .$$

Applying these transforms to the first term in (F.1) gives

$$
\begin{aligned}
a_2' t \int \left[m\left(\boldsymbol{r}\right)\right]^2 \mathrm{d}^3 r &= a_2' t \int \mathrm{d}^3 r \frac{1}{\left(2\pi\right)^6} \int m\left(\boldsymbol{q}\right) \exp\left(\mathrm{i}\boldsymbol{q}\boldsymbol{r}\right) \mathrm{d}^3 q \\
&\quad \times \int m^*\left(\boldsymbol{q}'\right) \exp\left(-\mathrm{i}\boldsymbol{q}'\boldsymbol{r}\right) \mathrm{d}^3 q' \\
&= a_2' t \frac{1}{\left(2\pi\right)^6} \int m\left(\boldsymbol{q}\right) m^*\left(\boldsymbol{q}'\right) \mathrm{d}^3 q\, \mathrm{d}^3 q' \\
&\quad \times \underbrace{\int \exp\left(\mathrm{i}\left(\boldsymbol{q} - \boldsymbol{q}'\right)\boldsymbol{r}\right) \mathrm{d}^3 r}_{\left(2\pi\right)^3 \delta\left(\boldsymbol{q} - \boldsymbol{q}'\right)} \\
&= a_2' t \frac{1}{\left(2\pi\right)^3} \int \left|m\left(\boldsymbol{q}\right)\right|^2 \mathrm{d}^3 q \quad , \tag{F.3}
\end{aligned}
$$

where the fact that since $m\left(\boldsymbol{r}\right)$ is real, it follows that $m\left(\boldsymbol{q}\right) = m^*\left(-\boldsymbol{q}\right)$ was used. Also note that in the third line of (F.3) the order of integration has been exchanged which allows one to derive the respective delta function $\delta\left(\boldsymbol{q} - \boldsymbol{q}'\right)$. The second term in (F.1) gives

$$
\begin{aligned}
g \int \left[\nabla m\left(\boldsymbol{r}\right)\right]^2 \mathrm{d}^3 r &= g \int \mathrm{d}^3 r \frac{1}{\left(2\pi\right)^6} \nabla \int m\left(\boldsymbol{q}\right) \exp\left(\mathrm{i}\boldsymbol{q}\boldsymbol{r}\right) \mathrm{d}^3 q \\
&\quad \times \nabla \int m^*\left(\boldsymbol{q}'\right) \exp\left(-\mathrm{i}\boldsymbol{q}'\boldsymbol{r}\right) \mathrm{d}^3 q' \\
&= g \int \mathrm{d}^3 r \frac{1}{\left(2\pi\right)^6} \int m\left(\boldsymbol{q}\right) \left(\mathrm{i}\boldsymbol{q}\right) \exp\left(\mathrm{i}\boldsymbol{q}\boldsymbol{r}\right) \mathrm{d}^3 q \\
&\quad \times \int m^*\left(\boldsymbol{q}'\right) \left(-\mathrm{i}\boldsymbol{q}'\right) \exp\left(-\mathrm{i}\boldsymbol{q}'\boldsymbol{r}\right) \mathrm{d}^3 q' \\
&= g \frac{1}{\left(2\pi\right)^6} \int m\left(\boldsymbol{q}\right) m^*\left(\boldsymbol{q}'\right) \left(\boldsymbol{q}\boldsymbol{q}'\right) \mathrm{d}^3 q\, \mathrm{d}^3 q' \\
&\quad \times \underbrace{\int \exp\left(\mathrm{i}\left(\boldsymbol{q} - \boldsymbol{q}'\right)\boldsymbol{r}\right) \mathrm{d}^3 r}_{\left(2\pi\right)^3 \delta\left(\boldsymbol{q} - \boldsymbol{q}'\right)} \\
&= g \frac{1}{\left(2\pi\right)^3} \int \boldsymbol{q}^2 \left|m\left(\boldsymbol{q}\right)\right|^2 \mathrm{d}^3 q \quad , \tag{F.4}
\end{aligned}
$$

which allows one to write the free energy as

$$F - F_0 = \frac{1}{\left(2\pi\right)^3} \int \mathrm{d}^3 q \left(a_2' t + g\boldsymbol{q}^2\right) \left|m\left(\boldsymbol{q}\right)\right|^2 \quad . \tag{F.5}$$

The result should now be compared to the solution for the Heisenberg model (7.26) where a similar k^2 dependence occurs. However, it must be kept in mind that the present case does not describe interacting spins, but local changes in the magnetization which in are represented by a superposition of plane waves in k space.

G. Bogoliubov–Peierls–Feynman Inequality

A Bogoliubov type inequality, which is used for the approximate calculation of the spin-fluctuation partition function, was first suggested by Peierls [179] in 1934. An elegant formal proof of this relation was given by Feynman [180] only 20 years later which is reviewed here. What one has to proof is the inequality

$$F \leq F_0 + \langle \mathcal{H} - \mathcal{H}_0 \rangle_0 \quad . \tag{G.1}$$

Here F is the exact free energy to the Hamiltonian H and F_0 is the free energy to a trial Hamiltonian $\mathcal{H}_0$. The inequality denotes that for any choice of $\mathcal{H}_0$ the exact free energy F represents a lower bound. This is exactly the condition which is known as the variational principle in quantum mechanics, the Bogoliubov inequality indeed represents a variational derivation of a mean-field theory.

The partition function should be

$$Z = \sum \exp(-\beta \mathcal{H}) \quad , \tag{G.2}$$

and the Hamiltonian should be broken up into

$$\mathcal{H} = \mathcal{H}_0 + \mathcal{H}_1 \quad , \tag{G.3}$$

where $\mathcal{H}_0$ should be chosen such that its respective partition function can be evaluated. One now considers the ratio of the true partition function to that of the trial Hamiltonian $\mathcal{H}_0$

$$\begin{aligned}
\frac{Z}{Z_0} &= \frac{\sum \exp(-\beta(\mathcal{H}_0 + \mathcal{H}_1))}{\sum \exp(-\beta \mathcal{H}_0)} \\
&= \sum P_0 \exp(-\beta \mathcal{H}_1) \\
&= \langle \exp(-\beta \mathcal{H}_1) \rangle_0 \quad . \tag{G.4}
\end{aligned}$$

P_0 is the correctly normalized Boltzmann probability factor for the system described by $\mathcal{H}_0$

$$P_0 = \frac{\exp(-\beta \mathcal{H}_0)}{Z_0} \quad , \tag{G.5}$$

and $\langle ... \rangle_0$ represents the average with respect to P_0.

Since for real arguments the first and second derivatives of the exponential function are always positive for any function f the relation holds

$$\langle \exp(f) \rangle \geq \exp \langle f \rangle \quad . \tag{G.6}$$

This relation is known as the convexity inequality (see Sect. A.) and describes a basic property of the free energy of the system. Although this relation must not hold locally everywhere it always holds globally since for reasons of the stability of a thermodynamical system, the free energy $f(x)$ must adopt an upward curvature at some stage when x goes to infinity. Applying this result one finds

$$\frac{Z}{Z_0} \geq \exp\left(-\beta \langle \mathcal{H}_1 \rangle_0\right) \quad . \tag{G.7}$$

Taking logarithms of both sides yields

$$\ln Z - \ln Z_0 \geq -\beta \langle \mathcal{H}_1 \rangle_0 \quad . \tag{G.8}$$

With the definition for the free energy

$$F = -k_B T \ln Z \quad , \tag{G.9}$$

and together with (G.3) one directly obtains the Bogoliubov inequality as defined in (G.1).

The Bogoliubov inequality is useful, because if one chooses $\mathcal{H}_0$ to be some simple approximation to the real Hamiltonian of a system, it represents a rigorous upper bound for the free energy of the true Hamiltonian. However, in order to obtain good results one will try to incorporate as much as possible of the physics, while still leaving the calculation tractable. One choice for $\mathcal{H}_0$ which is always possible is one that decomposes into a sum of many terms, one for each of the problem's degrees of freedom. But this is precisely the idea of the mean-field theory namely to replace the actual interaction between the parts of the system by a fictitious interaction with some external field or potential [181].

H. The Factor 2 in Equation (7.27)

One starts from

$$\gamma_k = \frac{1}{z} \sum_{\delta} \cos(k\delta) \simeq \frac{1}{z} \sum_{\delta} \left(1 - \frac{(k\delta)^2}{2} + \dots\right) , \tag{H.1}$$

where k is a wave vector of the magnon and δ is the position vector of the nearest-neighbor atom and will thus be named R_n. Since R_n describes a point of the lattice it can be written in terms of the lattice vectors (a_1, a_2, a_3)

$$R_n = n_1 a_1 + n_2 a_2 + n_3 a_3 \quad . \tag{H.2}$$

The possible magnon wave vectors are determined by the reciprocal lattice vectors $\boldsymbol{b}_j$ so that

$$\boldsymbol{k} = \sum_{j=1}^{3} \kappa_j \boldsymbol{b}_j \quad , \tag{H.3}$$

with

$$\kappa_j = \frac{m_j}{N_j} \quad m_j \in \mathsf{N} \quad \text{and} \quad N_j = \frac{L}{|\boldsymbol{a}_j|} \quad ,$$

with L being the length of the crystal. Thus $\boldsymbol{k}$ can be written as

$$\boldsymbol{k} = \sum_{j=1}^{3} \frac{m_j}{N_j} \boldsymbol{b}_j \quad . \tag{H.4}$$

The product $\boldsymbol{k\delta}$ can now be written as

$$\boldsymbol{k\delta} = \sum_{j=1}^{3} \sum_{i=1}^{3} \frac{m_j n_i}{N_j} \underbrace{\boldsymbol{a}_i \boldsymbol{b}_j}_{2\pi\delta_{ij}}$$

$$= 2\pi \sum_{i=1}^{3} \frac{m_i n_i}{N_i} \quad , \tag{H.5}$$

where the general relation between the direct an reciprocal lattice $\boldsymbol{a}_i \boldsymbol{b}_j = 2\pi\delta_{ij}$ was used. The square of $\boldsymbol{k\delta}$ is then given by

$$\sum_{\delta} (\boldsymbol{k\delta})^2 = (2\pi)^2 \sum_{(n_1,n_2,n_3)} \left(\sum_{i=1}^{3} \frac{m_i n_i}{N_i} \right)^2 \quad ,$$

where (n_1, n_2, n_3) are the triples describing the 6 nearest-neighbor positions in a simple cubic lattice. These are

$$(n_1, n_2, n_3) = \begin{cases} \pm 1, 0, 0 \\ 0, \pm 1, 0 \\ 0, 0, \pm 1 \end{cases} \quad .$$

Thus carrying out the summation over the nearest-neighbor sites one gets

$$\sum_{\delta} (\boldsymbol{k\delta})^2 = (2\pi)^2 \, 2 \sum_{i} \left(\frac{m_i}{N_i} \right)^2$$

$$= 2 \, (2\pi)^2 \frac{k^2}{b^2} \quad , \tag{H.6}$$

where the last term of (H.6) is due to the fact that

$$k^2 = \sum_{j=1}^{3} \left(\frac{m_j}{N_j} \right)^2 b_j^2 = b^2 \sum_{j=1}^{3} \left(\frac{m_j}{N_j} \right)^2 \quad ,$$

for $b = b_1 = b_2 = b_3$ for a cubic lattice. Since in turn $b = \frac{2\pi}{a}$ one obtains

$$\sum_\delta (k\delta)^2 = 2a^2 k^2 \quad , \tag{H.7}$$

so that

$$\gamma_k = \frac{1}{z} \sum_\delta \cos(k\delta)$$

$$\simeq \frac{1}{z} \sum_\delta \left(1 - \frac{(k\delta)^2}{2} + \dots \right)$$

$$\simeq 1 - \frac{1}{2z} \sum_\delta (k\delta)^2 = 1 - \frac{a^2 k^2}{z} \quad . \tag{H.8}$$

I. Hund's Rules

For systems with localized electrons the sequence of occupation of the electronic states is governed by Hund's rules which read:

1. For a given configuration the term with the maximum multiplicity is the one with the lowest energy.
2. For given configuration and multiplicity the term with the largest value of the angular momentum is the one with the lowest energy.
3. For given configuration, multiplicity and angular momentum, the term with the lowest value of the total momentum J is the one with the lowest energy if the configuration represents a less than half filled shell. If the shell is more than half filled, the term with the largest J is the lowest in energy.

How these rules work is best explained using an example e.g. vanadium. The term-symbol represents the angular momentum and spin state of a system consisting of a finite number of electrons. The angular momentum L and the total spin S are given as

$$L = \sum_i m_i \quad , \quad S = \sum_i s_i \quad . \tag{I.1}$$

The total angular momentum J within the Russel–Saunders coupling is given as

$$J = L \pm |S| \quad . \tag{I.2}$$

The degeneracy of the ground state is given by the multiplicity which reads $2|S| + 1$. In symbolic form the term symbol is written

$$^{(2|S|+1)} L_{(|J|)} \quad .$$

Vanadium has the configuration $[Ar],4s^2, 3d^3$. This means the fully occupied core levels have the same configuration as the noble gas Ar. These full shells do not contribute to the properties of the free ion and are thus disregarded for the term symbol. Also the 4s-shell is filled with two electrons and can also be neglected. The term symbol is thus determined from the not fully occupied shells only. The first rule demands to maximize the multiplicity, meaning that all three 3 electrons must be either spin-up or spin-down giving $|S| = \frac{3}{2}$. The second demands to maximize the angular momentum. Since the state with the largest value of m is the one with the lowest energy, one starts with the occupation of the $m = 2$ state, so that L takes the value 3. Similar to the one-electron states one uses letters to symbolize the respective L value so that $L = 0, 1, 2, 3, ...$ is written as $S, P, D, F, ...$ (notice that in contrast to the single-electron states capital letters are used). The last rule finally tells how to distinguish between $L + |S|$ and $L - |S|$ (depending on the sign of S). In the present case the shell is less than half filled so that $J = L - |S| = 3 - \frac{3}{2} = \frac{3}{2}$ gives the total angular momentum. The table below shows the resulting occupation of the three of the ten possible $3d$ states in Vanadium:

m	2	1	0	-1	-2
$s = \frac{1}{2}$	-	-	-	-	-
$s = -\frac{1}{2}$	↓	↓	↓	-	-

In the form of a term symbol this is written as

$$^4F_{3/2}$$

The total magnetic moment of such an atom consists of an orbital and a spin contribution

$$\mu_l = g_l \mu_B \frac{L}{\hbar} \quad , \quad \mu_S = g_s \mu_B \frac{S}{\hbar} \quad , \tag{I.3}$$

where $g_l = -1$ and $g_s = -2$ are the respective gyromagnetic factors. Due to the spin-magnetic anomaly ($g_s = -2$) the total magnetic moment is not simply the sum of the angular and the spin component (given by J) but is given by multiplication by the Landé factor g_j

$$\mu_j = g_j \mu_B \frac{J}{\hbar} \quad , \text{with} \tag{I.4}$$

$$g_j = -\frac{3J(J+1) - L(L+1) + S(S+1)}{2J(J+1)} \quad . \tag{I.5}$$

While in the present case both the angular- and the spin-momentum would give a magnetic moment of $3\mu_B$, due to the antiparallel coupling, the total moment becomes only $0.6\mu_B$.

The properties of the free atom configuration of the transition metals are tabulated below. Applying Hund's rules the determination is straight forward with two exceptions namely Cr and Cu (denoted by asterisks). In these two atoms there exists more than one partly filled shell. One thus can construct two term symbols which are determined from either the sum or the difference of the total spin within the two shells. Hund's rules do not account for this case, so that there exists no simple solutions which of the two term symbols describes the actual ground state of the atom. However, in most cases the state with an antiparallel coupling of the totals spins is the one with the lowest energy [182].

Table A.1. Term symbols of the 3d-transition metals according to Hund's rules

	configuration	L	S	J	g_j	$\mu_j[\mu_B]$	termsymbol
Sc	$[Ar],4s^23d^1$	$\frac{4}{2}$	$\frac{1}{2}$	$\frac{3}{2}$	-0.8	-1.2	$^2D_{3/2}$
Ti	$[Ar],4s^23d^2$	$\frac{6}{2}$	$\frac{2}{2}$	$\frac{4}{2}$	-0.67	$-1.3\dot{3}$	$^3F_{\frac{4}{2}}$
V	$[Ar],4s^23d^3$	$\frac{6}{2}$	$\frac{3}{2}$	$\frac{3}{2}$	-0.40	-0.60	$^4F_{3/2}$
Cr*	$[Ar],4s^13d^5$	$0\,\lvert 0$	$\frac{1}{2}\,\lvert\frac{5}{2}$	$\frac{1}{2}\,\lvert\frac{5}{2}$	$-2\,\lvert-2$	$-1\,\lvert-5$	$^2S_{1/2}\,\lvert^6S_{5/2}$
Mn	$[Ar],4s^23d^5$	0	$\frac{5}{2}$	$\frac{5}{2}$	-2	-5	$^6S_{5/2}$
Fe	$[Ar],4s^23d^6$	$\frac{4}{2}$	$\frac{4}{2}$	$\frac{8}{2}$	-1.5	-6.0	$^5D_{8/2}$
Co	$[Ar],4s^23d^7$	$\frac{6}{2}$	$\frac{3}{2}$	$\frac{9}{2}$	-1.33	-6.0	$^4F_{9/2}$
Ni	$[Ar],4s^23d^8$	$\frac{6}{2}$	$\frac{2}{2}$	$\frac{8}{2}$	-1.25	-5.0	$^3F_{8/2}$
Cu*	$[Ar],4s^13d^{10}$	$0\,\lvert 0$	$\frac{1}{2}\,\lvert 0$	$\frac{1}{2}\,\lvert 0$	$-2\,\lvert 0$	$-1\,\lvert 0$	$^2S_{1/2}\,\lvert^1S_0$
Zn	$[Ar],4s^23d^{10}$	0	0	0	0	0	$^1S_0$

J. Polynomial Coefficients in (18.12)

The terms which appear in the expansion of the free energy 18.12 are of the general form

$$F_{mtl}M^{2m}\left\langle m_\perp^2\right\rangle^t\left\langle m_\parallel^2\right\rangle^l \tag{J.1}$$

and F_{mtl} can be written as [213]

$$F_{mtl}=\frac{(m+t+l)!!2^t\,(2m+2l-1)!!}{m!l!\,(2m-1)!} \tag{J.2}$$

For a given order of the polynomial n, the coefficients F_{mtl} of the single terms can be taken form the tables below where the additional condition has to be obeyed: $m = n - t - l$

n=1	$l=0$	$l=1$
$t=0$	1	1
$t=1$	2	

n=2	$l=0$	$l=1$	$l=2$
$t=0$	1	6	3
$t=1$	4	4	
$t=2$	8		

n=3	$l=0$	$l=1$	$l=2$	$l=3$
$t=0$	1	15	45	15
$t=1$	6	36	18	
$t=2$	24	24		
$t=3$	48			

n=4	$l=0$	$l=1$	$l=2$	$l=3$	$l=4$
$t=0$	1	28	210	420	105
$t=1$	8	120	360	120	
$t=2$	48	288	144		
$t=3$	192	192			
$t=4$	348				

n=5	$l=0$	$l=1$	$l=2$	$l=3$	$l=4$	$l=5$
$t=0$	1	45	630	3150	4725	945
$t=1$	10	280	2100	4200	1050	
$t=2$	80	1200	3600	1200		
$t=3$	480	2880	1440			
$t=4$	1920	1920				
$t=5$	3840					

n=6	$l=0$	$l=1$	$l=2$	$l=3$	$l=4$	$l=5$	$l=6$
$t=0$	1	66	1485	13860	51975	62370	10395
$t=1$	12	540	7560	37800	56700	11340	
$t=2$	120	3360	25200	50400	12600		
$t=3$	960	14400	43200	14400			
$t=4$	5760	34560	17280				
$t=5$	23040	23040					
$t=6$	46080						

K. Conversion Between Magnetic Units

Theoreticians and experimentalists use often different units. While theoreticians prefer atomic units, experimentalists prefer to make use of Gaussian or SI units. Table A.2 provides help in converting between these different worlds. In the Gaussian system the flux density B and the field strength H are given in units of Gauss [G] and Oerstedt [Oe] respectively. Both units have the same numerical value so that the magnetization M can be given either in [G] or in [Oe]. The SI (System International) makes a distinction by measuring the flux density in Tesla [T] and the field strength and the magnetization in [A/m]. The conversion between B and H in the SI system is provided by $B = \mu_0 H$ where $\mu_0 = 4\pi 10^{-7} [\text{Tm/A}]$. In the atomic system, the units are derived from the properties of the hydrogen atom. The unit of length is the Bohr radius of the 1s orbital. The unit of energy equals twice the ionization energy of the H atom (1 Hartree) so that 1 [Ry] is exactly the ionization energy.

Table A.2. Useful conversions between atomic-, Gaussian-, and SI-units

	atomic units	Gaussian units (CGS)	International units (SI)
length	1 Bohr$= \frac{\hbar^2}{me^2}$	5.2918×10^{-9} cm	5.2918×10^{-11} m
mass	1 (electron restmass)	9.1096×10^{-28} g	9.1096×10^{-31} kg
charge	1 (electron charge)	4.8029×10^{-10} esu	1.6022×10^{-19} As
energy	1Hartree$=$2Ry$=$27, 208 eV	4.3592×10^{-11} erg	4.3592×10^{-18} J
Planck's constant	$\hbar = \frac{h}{2\pi} = 1$	1.0546×10^{-27} erg s	1.0546×10^{-34} J s
speed of light	1	$2.9979 \times 10^{+10}$ cm s^{-1}	$2.9979 \times 10^{+8}$ m s^{-1}
Boltzmann's const.	$k_B = 3.1671 \times 10^{-6}$ Hartree K^{-1}	1.3806×10^{-16} erg K^{-1}	1.3806×10^{-23} J K^{-1}
Bohr's magneton	$1\mu_B = \frac{eh}{2mc} = \frac{1}{2} \frac{\mu_B^2}{}$	9.2741×10^{-21} erg G^{-1}	9.2741×10^{-24} J T^{-1}
susceptibility	$\chi = \frac{M}{H} = 1 \frac{}{\text{Hartree atom}}$	$8.4149 \times 10^{+6}$ erg mol G^{-2}	$8.4149 \times 10^{+7}$ J mol T^{-2}
pressure	$P = \frac{E}{V} = 1 \frac{\text{Hartree}}{\text{bohr}^3}$	3.3994×10^{-15} erg cm^{-3}	3.3994×10^{-14} Pa

Further useful relations:

- energy: 1 eV$= 1.6022 \times 10^{-12}$ erg$= 1.6022 \times 10^{-19}$ J
- Bohr's magneton: 1 $\mu_B = 5.7884 \times 10^{-5}$ eV/T$= 4.2588 \times 10^{-6}$ Ry/T
- pressure: 1GPa$= 10$ kbar, P$[\frac{\text{Ry}}{\text{bohr}^3}] \times 1.47036 \times 10^{+5} = $ P[kbar]
- the unit 1 erg G^{-1} is identical to 1 emu

References

1. J. Hirschenberger, *Geschichte der Philosophie*, Vol.1 p.19, (Verlag Herder, Freiburg im Breisgau, 1980)
2. J. Needham, *Science in Civilisation in China*, vol. 4, Cambridge University Press (1962).
3. A. Neckam, *Libri II de naturis rerum*, reprint London 1863 (Wright Ed.).
4. J.S.T. Gehler, in *Physikalisches Wörterbuch* Vol. 6.2., p. 952, (Leipzig 1836).
5. P. Peregrinus, in his work from 1538, Peregrinus refers to the original Epistola... from 1269. Reprinted in: *Neudrucke von Schriften und Karten über Meteorologie und Erdmagnetismus*, ed. G. Hellmann (Berlin 1898).
6. W. Gilbert, *De Magnete*, facsimile reprint, Berlin 1892.
7. G. Knight, Phil. Trans. Roy. Soc. London **69** 161 (1744), ibid. **44** 656 (1747).
8. L. da Ponte, textbook to W.A. Mozarts opera Cosi fan tutte, (Vienna, 1795).
9. J.H. van Vleck, Intern. J. Magnetism **1** 1 (1970).
10. W. Sturgeon, Trans. Soc. for the Encouragement af Arts, Manufacures and Commerce **43** 37 (1826).
11. P. Curie, Ann. de Chim. et Phys. **5** 289 (1895).
12. P. Langevin, J. de Physique, **4** 678 (1905).
13. W. Pauli, Zeitsch. f. Physik **2** 201 (1920).
14. J.H. van Leeuwen, Dissertation, Leiden (1919) and J. de Physique **2** 361 (1921).
15. J.H. van Vleck, *The theory of electric and magnetic susceptibilities*, Oxford University Press (1932).
16. P. Weiss, J. de Physique **6** 661 (1907).
17. E. Landé , Zeitsch. f. Physik **15** 189 (1923).
18. G.E. Uhlenbeck and S. Goudsmit, Naturwiss. **13** 953 (1925), and Nature **117** 264 (1926).
19. P.A.M. Dirac, Proc. Roy. Soc. **117A** 610 (1928), ibid. **118A** 351 (1928).
20. F. Hund, *Linienspektren und Periodisches System der Elemente*, (Julius Springer, Berlin 1927).
21. W. Heitler and F. London, Zeitsch. f. Physik **44** 455 (1927).
22. W. Heisenberg, Zeitsch. f. Physik **38** 441 (1926).
23. P.A.M. Dirac, Proc. Roy. Soc. A **112** 661 (1926).
24. L. Neél, Ann. Phys. **3** 137 (1948).
25. F. Bloch, Zeitsch. f. Physik **57** 545 (1929).
26. W. Pauli, Zeitsch. f. Physik **41** 81 (1927).
27. N.F. Mott, Proc. Phys. Soc. (London) **47** 571 (1935).
28. J.C. Slater, Phys. Rev. **49** 537 (1936).
29. J.C. Slater, Phys. Rev. **49** 931 (1936).
30. E.C. Stoner, Proc. Roy. Soc. A **154** 656 (1936), ibid. **165** 372 (1938).
31. E.P. Wohlfarth, Rev. Mod. Phys. **25** 211 (1953).
32. J.C. Slater, Phys. Rev. **81** 385 (1951).

222   References

33. K. Schwarz, Phys. Rev. B **5** 2466 (1972).
34. P. Hohenberg and W. Kohn, Phys. Rev. **136** 864 (1964).
35. W. Kohn and L.J. Sham, Phys. Rev. **140** 1133 (1965).
36. R. Gaspar, Act. Phys. Ac. Sci. Hung. **3** 263 (1954).
37. L. Hedin and B. L. Lundquist, J. Phys. C: Condens. Matter **4** 2064 (1971).
38. U. von Barth and L. Hedin, J. Phys. C: Condens. Matter **5** 1629 (1972).
39. E.P. Wohlfarth, J. Appl. Phys. **41** 1205 (1970).
40. S. Wakoh and J. Yamashita, J. Phys. Soc. Japan **28** 115 (1970).
41. A.M. Oles and G. Stollhoff, J. Magn. Magn. Mater. **54-57** 1045 (1986).
42. J. Hubbard, Proc. Roc. Soc. A **276** 238 (1963).
43. T. Moriya in *Spin Fluctuations in Itinerant Electron Magnetism*, Springer Ser. Solid - State Sci., Vol. 56 (Springer, Berlin, Heidelberg 1985).
44. K.K. Murata and S. Doniach, Phys. Rev. Letters **29** 285 (1972).
45. G.G. Lonzarich, J. Magn. Magn. Mater. **45** 43 (1984).
46. G.G. Lonzarich and J. Taillefer, J. Phys. C: Condens. Matter **18** 4339 (1985).
47. P. Mohn and E.P. Wohlfarth, J. Phys. F: Metal Physics **17** 2421 (1986).
48. L.D. Landau, Z. Phys. **64** 629 (1930).
49. R. Peierls, Z. Phys. **80** 763 (1933).
50. A.H. Wilson, Proc. Cambridge Philos. Soc. **49** 292 (1953).
51. R.T. Schumacher, C. P. Slichter, Phys. Rev. **101** 58 (1956).
52. R.T. Schumacher and W.E. Vehse, J. Phys. Chem. Solids **24** 297 (1965).
53. S. Schultz and G. Dunifer, Phys. Rev. Letters **18** 283 (1967).
54. J.A. Kaeck, Phys. Rev. **175** 897 (1968).
55. E. Burzo, Int. J. Magn. **3** 161 (1972).
56. R. Herrmann, U. Preppernau *Elektronen im Kristall* (Springer-Verlag, Wien, New York 1979).
57. R.M. Bozorth, *Ferromagnetism* (Van Nostrand, New York, 1951).
58. C. J. Nieuwenhuys, Adv. Phys. **24** 515 (1975)
59. C.E. Guillaume, CR Acad, Sci. **125** 235 (1897).
60. C.J. Davisson and L. H. Germer, Nature **119** 558 (1927).
61. C.J. Davisson and L.H. Germer, Phys. Rev. **30** 705 (1927).
62. G.F. Bacon, *Neutron Diffraction*, 2nd. ed. (Oxford University Press, Oxford 1962).
63. H. Akai, M. Akai, S. Blügel, B. Drittler, H. Ebert, K. Terakura, R. Zeller, and P.H. Dederichs, Progr. Theor. Phys. **101** 1 (1990).
64. P. Mohn, Hyperfine Interactions **128** 67 (2000).
65. G. Schütz, W. Wagner, W. Wilhelm, P. Kienle, R. Zeller, R. Frahm, and G. Materlik, Phys. Rev. Letters **58** 737 (1987).
66. K. Schwarz, J. Phys. F: Metal Physics **16** L211 (1986).
67. G. Binnig, H. Rohrer, Scientific American, **253** 40 (1985).
68. G. Binnig, H. Rohrer, Reviews of Modern Physics, **71** S324 (1999).
69. G. Binnig, H. Rohrer, IBM J. Res. Develop. **30** 355 (1986).
70. W. J. de Haas, P. M. van Alphen, Leiden Comm. **208**d, **212**a (1930).
71. B. Lengeler, *Springer Tracts Mod. Phys.* **82**, 1 (Springer, Berlin, Heidelberg, New York 1978).
72. M. Cardona and L. Ley (eds.) *Photoemission in Solids I, II*, Topics in Applied Physics, **26**, **27** (Springer, Berlin, Heidelberg, New York 1979).
73. B. Feuerbacher, B. Fitton, and R.F. Willis (eds.) *Photoemission and the Electronic Properties of Surfaces*, (John Wiley and Sons, New York 1978).
74. W.E. Henry, Phys. Rev. **88** 559 (1952).
75. H.A. Brown and J.M. Luttinger, Phys. Rev. **100** 685 (1955).
76. P.R. Rhodes and E.P. Wohlfarth, Proc. Roy. Soc. **273** 247 (1963).

77. C. Kittel, *Quantum Theory of Solids* (John Wiley and Sons, New York, Chicester, Brisbane, Toronto, Singapore, 1987).
78. T. Holstein and H. Primakoff, Phys. Rev. **58** 1908 (1940).
79. F.J. Dyson Phys. Rev. **102** 1217 (1956).
80. F. Keffer and R. Loudon J. Appl. Phys. (Suppl.) **32** 2 (1961).
81. F. Bloch, Z. Physik **61** 206 (1930).
82. E. Ising, Z. Phys. **31** 253 (1925).
83. J.M. Yeomans, *Statistical Mechanics of Phase Transitions*, Oxford Science Publications, Clarendon Press, Oxford 1992.
84. L. Onsager, Phys. Rev. **65** 117 (1944).
85. D. Mermin and H. Wagner, Phys. Rev. Letters **17** 1133 (1966).
86. E.P. Wohlfarth, Phys. Lett. **3** 17 (1962).
87. R. Gersdorf, J. Phys. Radium **23** 726 (1962).
88. J.F. Janak, Phys. Rev. B **16** 255 (1977).
89. J. Korringa, Physica **13** 392 (1947).
90. W. Kohn and N. Rostoker, Phys. Rev. **94** 1111 (1954).
91. E.P. Wohlfarth and P. Rhodes, Phil. Mag. **7** 1817 (1962).
92. K. Schwarz and P. Mohn, J. Phys. F: Metal Physics **14** L128 (1984).
93. C.J. Schinkel, J. Phys. F: Met. Phys. **8** L87 (1978).
94. R. Lemaire, Cobalt **33** 201 (1966).
95. T. Goto, T. Sakakibara, K. Murata, and H. Komatsu, Solid State Commun. **12** 945 (1989).
96. F. Laves and H. Witte, Metallwirtsch. **15** 840 (1936).
97. T. Jarlborg and A.J. Freeman, Phys. Rev. B **22** 2342 (1980).
98. B.T. Matthias, A.L. Giorgi, V.O. Struebing, and J.L. Smith, Phys. Lett. A **69** 221 (1978).
99. B.T. Matthias and R.M. Bozorth, Phys. Rev. **109** 604 (1958).
100. V.V. Aleksandryan, A.S. Laugtin, R.Z. Levitin, A.S. Markosyan, and V.V. Snegirev Soy. Phys.-JETP **62** 153 (1985).
101. K. Ishiyama, K. Endo, T. Sakakibara, T. Goto, K. Sugiyama, and M. Date, J. Phys. Soc. Japan **56** 29 (1987).
102. M. Iijima, K. Endo, T. Sakakibara, and T. Goto, J. Phys.: Condens. Matter **2** 10069 (1990).
103. P. Mohn and K. Schwarz, Solid State Commun. **57** 103 (1986).
104. J. Abart and J. Voitländer, Solid State Commun. **40** 227 (1981).
105. U. Gradmann and H.O. Isbert J. Magn. Magn. Mater. **15** 1109 (1980).
106. D. Pescia, M. Stampanoni, G.L. Bona, A. Vaterlaus, R.F. Willis, and F. Weier, Phys. Rev. Letters **58** 2126 (1987).
107. Y. Tsunoda, J. Phys.: Condens. Matter **1** 10427 (1989).
108. A.P. Malozemoff and A.R. Williams, V. L. Moruzzi, Phys. Rev. B **29** 1620 (1984).
109. J. Friedel, Nouvo Cimento **10**, Suppl. 2 287 (1958).
110. J.L. Calais and K. Schwarz, Israel J. Chem. **19** 88 (1980).
111. A.R. Williams, V.L. Moruzzi, A.P. Malozemoff, and K. Terakura, *IEEE Transactions on Magnetism* **MAG-19**, 1983 (1983).
112. P. Mohn and D.G. Pettifor, J. Phys. C: Solid State Physics, **21** 2829 (1988), ibid.**21** 2841 (1988).
113. V. Heine, private communication.
114. D.G. Pettifor, Materials Science and Technology **4** 2480 (1988).
115. P. Söderlind, R. Ahuja, O. Eriksson, J.M. Wills, and B. Johansson, Phys. Rev. B **50** 5918 (1994).
116. A.R. Williams, R. Zeller, V.L. Moruzzi, Gelatt C.D.Jr. and J. Kübler, J. Appl. Phys. **52** 2067 (1981).

117. D.G. Pettifor, *Metallurgical Chemistry* (ed. O. Kubaschewski) p.191 (Her Majesty's Stationary Office, London, 1972).
118. P. Mohn, K. Schwarz, Physica B **130** 26 (1985).
119. K. Schwarz, P. Mohn, P. Blaha, and J. Kübler, J. Phys. F: Metal Physics **14** 2659 (1984).
120. A. Oswald, R. Zeller, and P. Dederichs, Phys. Rev. Letters **56** 1419 (1986).
121. J.F. van Acker, W. Speier, and R. Zeller, Phys. Rev. B **43** 9558 (1991).
122. P.W. Anderson, Phys. Rev. **124** 41 (1961).
123. P.A. Wolff, Phys. Rev. **124** 1030 (1961).
124. A.M. Clogston, Phys. Rev. **125** 439 (1961).
125. T. Moriya, Prog. Theor. Phys. **34** 329 (1965).
126. M.A. Rudermann and C. Kittel, Phys.Rev. **96** 99 (1954).
127. T. Kasuya, Progr. Theor. Phys. (Japan) **16** 45 (1965).
128. K. Yoshida, Phys. Rev. **106** 893 (1957).
129. P. Dederichs, *24. IFF-Ferienkurs: Magnetismus von Festkörpern und Grenzflächen*, Chap. 27 (Forschungszentrum Jülich, 1983).
130. D. Gignoux, R. Lemaire, P. Molho, and F. Tasset, J. Appl. Phys. **52** 2087 (1981).
131. M. Shimizu and J. Inoue, J. Phys. F: Metal Physics **17** 1221 (1987).
132. A.T. Aldred, Phys. Rev. B **11** 2597 (1975).
133. D.M. Edwards, E. P. Wohlfarth, Proc. Roy. Soc. A **303** 127 (1968).
134. E.D. Thompson, E.P. Wohlfarth and A.C. Bryan, Proc. Phys. Soc. **83** 59 (1964).
135. J. Hubbard, Phys. Rev. B **19** 2626 (1979).
136. J. Hubbard, Phys. Rev.B **20** 4584 (1979).
137. Y. Nagaoka, Phys. Rev. **147** 392 (1966).
138. N.F. Mott, Rev. Mod. Phys. **40** 677 (1968).
139. N.F. Mott, *Metal-Insulator Transitions* (Taylor and Francis, London, 1974).
140. M.C. Gutzwiller, Phys. Rev. **137** 1726 (1965).
141. H.A. Razafimandimby, Z. Phys.: Condens. Matter **49** 33 (1982).
142. K. Baberschke, M. Donath, and W. Nolting (Eds.), *Band-Ferromagnetism; Ground-State and Finite-Temperature Phenomena*, Lecture Notes in Physics, LNP580, (Springer Verlag, 2001).
143. J.-C. Tolédano and P. Tolédano, *The Landau Theory of Physe Transitions*, (World Scientific, Singapore, 1987).
144. E. Burzo, A. Chelkowski, and H.R. Kirchmayr *Compounds Between Rare Earth Elements and 3d, 4d or 5d Eelements*, Landolt-Börnstein, New Series, Group III: Crystal and Solid State Physics, Vol. 19, subvolume d2, (Springer Verlag, 1990).
145. D. Bloch, D.M. Edwards, and M. Shimizu, J. Voiron, J. Phys. F: Metal Phys. **5** 1217 (1975).
146. J. Kübler, *Theory of Itinerant Electron Magnetism*, International Series of Monographs in Physics, (Clarendon Press, Oxford, 2000).
147. L.D. Landau, E.M. Lifschitz, *Lehrbuch der Theoretischen Physik* vol.5, p.318, (Akademie Verlag, Berlin, 1979).
148. Q. Qi, R. Skomski, and J.M. Coey, J. Phys.: Condens. Matter **6** 3245 (1994).
149. S.S. Jaswal, Phys. Rev. B **48** 6156 (1993).
150. B. Rellinghaus, J. Kästner, T. Schneider, E.F. Wassermann, and P. Mohn, Phys. Rev. B **51** 2983 (1995).
151. P. Mohn and G. Hilscher, Phys. Rev. B **40** 9126 (1989).
152. J.G.M. Armitage, R.G. Graham, P.C. Riedi, and J.S. Abell, J. Phys.: Condens. Matter **2** 8779 (1990).

153. E.F. Wassermann, *Ferromagnetic materials*, vol.5, eds. K. H. J. Buschow and E. P. Wohlfarth (North-Holland, Amsterdam, 1990).
154. R.J. Weiss, Proc. Roy. Soc. London **82** 281 (1963).
155. A.R. Williams, V.L. Moruzzi, C. D. Gelatt Jr., J. Kübler, and K. Schwarz, J. Appl. Phys. **53** 2019 (1982).
156. P. Mohn, K. Schwarz, and D. Wagner, Physica B **161** 139 (1989).
157. P. Mohn, K. Schwarz, and D. Wagner, Phys. Rev. **B 43** 3318 (1991).
158. P. Mohn and K. Schwarz, J. Magn. Magn. Mater. **104-107** 685 (1992).
159. P. Mohn and K. Schwarz, Int. J. Mod. Physics B **7** 579 (1993).
160. J. C. Ododo, J. Phys. F: Met. Phys. **15** 941 (1985).
161. P. Mohn and K. Schwarz, J. Phys.: Condens. Matter **5** 5099 (1993).
162. R.M. White, *Quantum theory of magnetism*, Springer Series in Solid-State Sciences, Vol. 32. , 2nd Ed. (Springer-Verlag, Berlin, Heidelberg, New York, 1983).
163. J. Lindhard, J. Kgl. Danske Videnskap. Selskap., Mat.-fys. Medd., **28** 8 (1954).
164. D. McKenzie Paul, P.W. Mitchell, H. A. Mook, and U. Steigenberger, Phys. Rev. **B 38** 580 (1988).
165. H.A. Mook and D. McKenzie Paul, Phys. Rev. Letters **54** 227 (1985).
166. E. Babic, Z. Maharonic, and E. P. Wohlfarth, Phys. Lett. **95A** 335 (1983).
167. D. Pines, and P. Nozières, *The theory of quantum fluids*, (W.A. Benjamin, New York, 1966).
168. D. Pines, *The many body problem*, (W.A. Benjamin, New York, 1962).
169. T. Izuyama, E.J. Kim, and R. Kubo, J. Phys. Soc. Japan **18** 1025 (1963).
170. C. Herring, *Magnetism* Vol. IV, G.T. Rado and H. Suhl eds. (Academic Press, New York, 1966).
171. J.A. Blackman, T. Morgan, and J.F. Cooke, Phys. Rev. Letters **55** 2814 (1985).
172. J.F. Cooke, J.A. Blackman, and T. Morgan, Phys. Rev. Letters **54** 718 (1985).
173. D.M. Edwards and R.B. Muniz, J. Phys. F: Metal Physics **15** 2357 (1985).
174. M.W. Stringfellow, J. Phys. C: Condens. Matter **1** 950 (1968).
175. A.T. Aldred and P.H. Froehle, Int. J. Magn. **2** 195 (1972).
176. O. Gunnarson, J. Phys. F: Metal Physics **6** 587 (1976).
177. K. Kirchner, W. Weber, J. Voitländer, J. Phys. Condes. Matter **4** 8097 (1992).
178. M. Tinkham, *Group theory and quantum mechanics*, (McGraw-Hill, New York 1964).
179. R. Peierls, Phys. Rev. **54** 918 (1934).
180. R.P. Feynman, Phys. Rev. **B97** 660 (1955).
181. J.J. Binney, N.J. Dowrick, A.J. Fisher, and M.E.J. Newman, *The theory of critical phenomena*, (Clarendon Press, Oxford, 1992).
182. E.U. Condon and G.H. Shortley, *The Theory of the Atomic Spectra*, (Cambridge University Press, 1963).
183. T. Moriya and Y. Takahashi, J. Phys. Soc. Japan **45** 397 (1978).
184. H. Hasegawa, J. Phys. Soc. Japan **46** 1504 (1979).
185. T. Moriya and K. Usami, Solid State Commun. **34** 95 (1980).
186. K. Usami and T. Moriya, J. Magn. Magn. Mater. **15-18** 847 (1980).
187. H. Hasegawa, J. Phys. Soc. Japan **49** 178 (1980).
188. H. Hasegawa, J. Phys. Soc. Japan **46** 963 (1980).
189. P. Entel and M. Schröter, J. Phys. France **49** C8-293 (1988).
190. P. Entel and M. Schröter, Physica B **161** 153 (1989).
191. D. Wagner, J. Phys.: Condens. Matter **1** 4635 (1989).
192. W. Nolting, A. Vega, and T. Fauster, Z. Phys. B **96** 357 (1995).
193. C. Herring, Phys. Rev. **85** 1003 (1952), ibid. **87** 60 (1952).

194. A. J. Pindor, J. Staunton, G.M. Stocks, and H. Winter, J. Phys. F: Metal Physics **13** 979 (1983).
195. J. Staunton, B.L. Gyorffy, A.J. Pindor, and G.M. Stocks, H. Winter, J. Magn. Magn. Mater. **45** 15 (1984).
196. B.L. Gyorffy, A.J. Pindor, J. Staunton, G.M. Stocks, and H. Winter, J. Phys. F: Metal Physics **15** 1337 (1985).
197. A.Z. Solontsov and D. Wagner, Phys. Rev. B **51** 12410 (1995).
198. R.E. Prange and V. Korenman, Phys. Rev. B **19** 4961 (1979).
199. R.E. Prange, V. Korenman, Phys. Rev. B **19** 4698 (1979).
200. H. Capellmann, Z. Phys. B **34** 29 (1979).
201. V. Heine, J.H. Samson, and C.M.M. Nex, J. Phys. F: Metal Phys. **11** 2645 (1981).
202. L.M. Sandratskii, phys. stat. sol. (b) **135** 167 (1986).
203. L.M. Sandratskii, P. G. Guletskii, J. Magn. Magn. Mater. 79 306 (1989).
204. L.M. Sandratskii, Solid State Commun. **75** 527 (1990).
205. L.M. Sandratskii, Adv. Phys. **47** 91 (1998).
206. M. Uhl, J. Kübler, Phys. Rev. Letters **77** 334 (1996).
207. M. Shimizu, Rep. Prog. Phys. **44** 329 (1981).
208. M. Uhl and J. Kübler, J. Phys.: Condens. Matter **9** 7885 (1997).
209. N.M. Rosengaard and B. Johansson, Phys. Rev. B **55** 14975 (1997).
210. S.V. Halilov, H. Eschrig, A.Y. Perlov, and P.M. Oppeneer, Phys. Rev. B **58** 293 (1998).
211. V.P. Antropov, M.I. Katsnelson, M. van Schilfgaarde, and B. N. Harmon, Phys. Rev. Letters **75** 334 (1995).
212. V.P. Antropov, M.I. Katsnelson, B.N. Harmon, M. van Schilfgaarde, and D. Kusnezov, Phys. Rev. B **54** 1019 (1996).
213. M. Uhl, *Spinfluktuationen; Thermodynamische Beschreibung Magnetischer Materialien*, (Dissertation, TU-Darmstadt, 1995).

Index

Springer Series in
SOLID-STATE SCIENCES

Springer Series in
SOLID-STATE SCIENCES

Series Editors:
M. Cardona   P. Fulde   K. von Klitzing   R. Merlin   H.-J. Queisser   H. Störmer